# 67 Advances in Polymer Science
Fortschritte der Hochpolymeren-Forschung

# Characterization of Polymers in the Solid State II:

Synchrotron Radiation, X-ray Scattering and Electron Microscopy

Editors: H. H. Kausch and H. G. Zachmann

With Contributions by
G. Bodor, G. Elsner, J. Hendrix, L. Monnerie,
C. Riekel, H. G. Stuhrmann, J.-L. Viovy,
I. G. Voigt-Martin, H. G. Zachmann

With 164 Figures and 10 Tables

Springer-Verlag
Berlin Heidelberg New York Tokyo

ISBN-3-540-13780-7 Springer-Verlag Berlin Heidelberg New York Tokyo
ISBN-0-387-13780-7 Springer-Verlag New York Heidelberg Berlin Tokyo

Library of Congress Catalog Card Number 61-642

This work is subject to copyright. All rights are reserved, whether the whole or part of the material is concerned, specifically those of translation, reprinting, re-use of illustrations, broadcasting, reproduction by photocopying machine or similar means, and storage in data banks. Under § 54 of the German Copyright Law where copies are made for other than private use, a fee is payable to the publisher, the amount to "Verwertungsgesellschaft Wort". Munich.

© Springer-Verlag Berlin Heidelberg 1985
Printed in Germany

The use of general descriptive names, trademarks, etc. in this publication, even if the former are not especially identified, is not to be taken as a sign that such names, as understood by the Trade Marks and Merchandise Marks Act. may accordingly be used freely by anyone

Typesetting: Th. Müntzer, GDR; Offsetprinting: Br. Hartmann, Berlin;
Bookbinding: Lüderitz & Bauer, Berlin

2154/3020-543210

# Preface

In the past decades much progress in the application of polymer materials was due to careful analysis of the interrelation between microstructure and macroscopic behavior of polymers. Any macroscopic information which is used, however, to characterize an inhomogeneous solid generally involves elemental response functions and orientation distributions which are not completely known. For this reason a certain ambiguity is introduced into the deconvolution of such experimental data. This leaves some room for individual interpretations as to the concentration and nature of structural elements and of defects and as to the mode of their interaction. Evidently, any analysis will be greatly facilitated if spatial and time resolution are improved and contrasts are created or increased. Exactly this was achieved in recent years by refining existing and meanwhile conventional techniques (IR- and X-ray analysis, electron microscopy) or by developing techniques towards new and special applications (deuteron NMR, cross-polarization magic angle spinning NMR, neutron scattering and synchrotron radiation). Several symposia and conferences have been organized on these topics. Examples are the EPS Conference on "New Developments in the Characterization of Polymers in the Solid State" in Hamburg, the "Symposium on Polymer Research at Synchrotron Radiation Sources" at Brookhaven, and the series of ACS Symposia on "Instrumental Methods for Polymer Characterization". These meetings testify two things: the notable progress achieved and the considerable interest in its application.

The editors feel that the topics selected for these two volumes on "Characterization" are *new* so as to qualify for this series, sufficiently *mature* so as to warrant a review article, and of timely interest to the readers. The editors are grateful to the authors for their willing collaboration and to the publisher for a speedy and careful production.

Lausanne and Hamburg, October 1984          H. H. Kausch
                                             H. G. Zachmann

# Editors

Prof. Hans-Joachim Cantow, Institut für Makromolekulare Chemie der Universität, Stefan-Meier-Str. 31, 7800 Freiburg i. Br., FRG

Prof. Gino Dall'Asta, Via Pusiano 30, 20137 Milano, Italy

Prof. Karel Dušek, Institute of Macromolecular Chemistry, Czechoslovak Academy of Sciences, 16206 Prague 616, ČSSR

Prof. John D. Ferry, Department of Chemistry, The University of Wisconsin, Madison, Wisconsin 53706, U.S.A.

Prof. Hiroshi Fujita, Department of Macromolecular Science, Osaka University, Toyonaka, Osaka, Japan

Prof. Manfred Gordon, Department of Pure Mathematics and Mathematical Statistics, University of Camebridge CB2 1SB, England

Prof. Gisela Henrici-Olivé, Chemical Department, University of California, San Diego, La Jolla, CA 92037, U.S.A.

Prof. Dr. habil. G. Heublein, Sektion Chemie, Friedrich-Schiller-Universität, Humboldtstraße 10, 69 Jena, DDR

Prof. Dr. H. Höcker, Universität Bayreuth, Makromolekulare Chemie I, Universitätsstr. 30, 8580 Bayreuth, FRG

Prof. Hans-Henning-Kausch, Laboratoire de Polymères, Ecole Polytechnique Fédérale de Lausanne, 32, ch. de Bellerive, 1007 Lausanne, CH

Prof. Joseph P. Kennedy, Institute of Polymer Science, The University of Akron, Akron, Ohio 44325, U.S.A.

Prof. Werner Kern, Institut für Organische Chemie der Universität, 6500 Mainz, FRG

Prof. Seizo Okamura, No. 24, Minamigoshi-Machi Okazaki, Sakyo-Ku. Kyoto 606, Japan

Prof. Salvador Olivé, Chemical Department, University of California, San Diego, La Jolla, CA 92037, U.S.A.

Prof. Charles G. Overberger, Department of Chemistry. The University of Michigan, Ann Arbor, Michigan 48 104, U.S.A.

Prof. Helmut Ringsdorf, Institut für Organische Chemie, Johannes-Gutenberg-Universität, J.-J.-Becher Weg 18–20, 6500 Mainz, FRG

Prof. Takeo Saegusa, Department of Synthetic Chemistry, Faculty of Engineering, Kyoto University, Kyoto, Japan

Prof. Günter Victor Schulz, Institut für Physikalische Chemie der Universität, 6500 Mainz, BRD

Prof. William P. Slichter, Chemical Physics Research Department, Bell Telephone Laboratories, Murray Hill, New Jersey 07971, U.S.A.

Prof. John K. Stille, Department of Chemistry. Colorado State University, Fort Collins, Colorado 80523, U.S.A.

# Editorial

With the publication of Vol. 51, the editors and the publisher would like to take this opportunity to thank authors and readers for their collaboration and their efforts to meet the scientific requirements of this series. We appreciate our authors concern for the progress of Polymer Science and we also welcome the advice and critical comments of our readers.

With the publication of Vol. 51 we should also like to refer to editorial policy: *this series publishes invited, critical review articles of new developments in all areas of Polymer Science in English (authors may naturally also include works of their own)*. The responsible editor, that means the editor who has invited the article, discusses the scope of the review with the author on the basis of a tentative outline which the author is asked to provide. Author and editor are responsible for the scientific quality of the contribution; the editor's name appears at the end of it.
Manuscripts must be submitted, in content, language and form satisfactory, to Springer-Verlag. Figures and formulas should be reproducible. To meet readers' wishes, the publisher adds to each volume a "volume index" which approximately characterizes the content.

Editors and publisher make all efforts to publish the manuscripts as rapidly as possible, i.e., at the maximum, six months after the submission of an accepted paper. This means that contributions from diverse areas of Polymer Science must occasionally be united in one volume. In such cases a "volume index" cannot meet all expectations, but will nevertheless provide more information than a mere volume number.

From Vol. 51 on, each volume contains a subject index.

Editors                                                                 Publisher

# Table of Contents

**Synchrotron Radiation in Polymer Science**
G. Elsner, Ch. Riekel, and H. G. Zachmann . . . . . . . .  1

**Position Sensitive X-ray Detectors**
J. Hendrix . . . . . . . . . . . . . . . . . . . . . 59

**Fluorescence Anisotropy Technique**
**Using Synchrotron Radiation as a Powerful Means for Studying the Orientation Correlation Functions of Polymer Chains**
J.-L. Viovy and L. Monnerie . . . . . . . . . . . . . . 99

**Resonance Scattering in Macromolecular Structure Research**
H. B. Stuhrmann . . . . . . . . . . . . . . . . . . 123

**X-ray Line Shape Analysis**
**A Means for the Characterization of Crystalline Polymers**
G. Bodor . . . . . . . . . . . . . . . . . . . . . 165

**Use of Transmission Electron Microscopy to Obtain Quantitative Information About Polymers**
I. G. Voigt-Martin . . . . . . . . . . . . . . . . . 195

**Author Index Volumes 1–67** . . . . . . . . . . . . 219

**Subject Index** . . . . . . . . . . . . . . . . . . . 229

# Synchrotron Radiation in Polymer Science

Gerhard Elsner*
Christian Riekel and Hans Gerhard Zachmann
Institut für Technische und Makromolekulare Chemie, Universität Hamburg,
D 2000 Hamburg 13, Martin-Luther-King Platz 6, FRG

*This article starts with a short introductory chapter on the generation and properties of synchrotron radiation. Afterwards instrumentation available for research on polymers are discussed and finally the results obtained up to now in polymer science.*

| | |
|---|---|
| **1 Introduction** | 3 |
| **2 Properties and Production of Synchrotron Radiation** | 3 |
|   2.1 How Does a Synchrotron Work? | 3 |
|   2.2 The Difference Between a Synchrotron and a Storage Ring | 7 |
|   2.3 Theory | 11 |
|   2.4 Comparison with Other Sources | 19 |
|   2.5 Present and Future Synchrotron Radiation Sources | 21 |
| **3 Instrumentation** | 22 |
|   3.1 Types of Synchrotron Radiation Small-Angle Scattering Cameras | 22 |
|     3.1.1 Energy Dispersive Cameras | 23 |
|     3.1.2 Angular Dispersive Cameras | 24 |
|   3.2 Optical Elements | 27 |
|     3.2.1 Monochromator | 27 |
|     3.2.2 Mirror | 27 |
|   3.3 Resolution | 28 |
|   3.4 Data Aquisition Systems | 31 |
|     3.4.1 One Dimensional Position Sensitive Detector | 31 |
|     3.4.2 Two Dimensional Position Sensitive Detector | 33 |
|   3.5 Ancillary Equipment | 34 |
|     3.5.1 Temperature Changes | 35 |
|     3.5.2 Stretching Experiments | 37 |
|     3.5.3 Chemical Reactions | 37 |
| **4 Results** | 38 |
|   4.1 Small Angle Scattering During Isothermal Crystallization | 38 |
|     4.1.1 Polyethylene Terephthalate | 38 |
|     4.1.2 Polyethylene | 41 |
|     4.1.3 Poly-β-hydroxybutyrate, PHB | 41 |

---

* Present address IBM Deutschland GmbH, PR NFS 0836, PO Box 266, 7032 Sindelfingen, FRG

4.2 Small Angle Scattering During Annealing Above the Crystallization Temperature . . . . . . . . . . . . . . . . . . . . . . . . . . . . . 41
    4.2.1 Polyethylene . . . . . . . . . . . . . . . . . . . . . . . . . 42
    4.2.2 Polyethylene Terephthalate . . . . . . . . . . . . . . . . 43
4.3 Small Angle Scattering During Phase Separations . . . . . . . . . 46
    4.3.1 Polymer Blends . . . . . . . . . . . . . . . . . . . . . . . 46
    4.3.2 Block Copolymers . . . . . . . . . . . . . . . . . . . . . 46
4.4 Small Angle Scattering During Crazing . . . . . . . . . . . . . . . 46
4.5 Wide Angle Scattering During Crystallization and Solid State Phase Transitions . . . . . . . . . . . . . . . . . . . . . . . . . . . . . . 47
    4.5.1 Polyisobutylene . . . . . . . . . . . . . . . . . . . . . . . 48
    4.5.2 Rubber . . . . . . . . . . . . . . . . . . . . . . . . . . . 48
    4.5.3 Polypropylene . . . . . . . . . . . . . . . . . . . . . . . . 49
    4.5.4 Polyamides . . . . . . . . . . . . . . . . . . . . . . . . . 50
    4.5.5 Polyethylene-naphthalene-2,6-dicarboxylate and Copolyesters . . . 51
    4.5.6 Polyethylene . . . . . . . . . . . . . . . . . . . . . . . . . 53
    4.5.7 Polyacetylene . . . . . . . . . . . . . . . . . . . . . . . . 53
4.6 Melting . . . . . . . . . . . . . . . . . . . . . . . . . . . . . . . . 54
4.7 Wide Angle Scattering During Chemical Reactions . . . . . . . . 54

**5 References** . . . . . . . . . . . . . . . . . . . . . . . . . . . . . . . . . 55

# 1 Introduction

The electrons or positrons which are moving on the circular orbit of the synchrotron emit a very strong electromagnetic radiation, the so-called synchrotron radiation. This effect was predicted first by Liénard [1] in 1898 and by Schott [2] in 1907. The complete theory was developed by Ivanenko and Pomeranchuck [3] in 1944 as well as by Schwinger [4] in 1946.

The first experimentalist who became interested in synchrotron radiation was Blewett [5] in 1946, who measured the energy loss due to this radiation. However, he was not able to detect the radiation itself. In 1947, according to the illustrative historical remarks made by Baldwin [6], F. Haber was able to see the synchrotron radiation with help of a mirror. The first investigations of the properties of this radiation were performed by Elder et al. [7]

Very soon one recognized that the synchrotron radiation was not only an unwanted by-product of the particle acceleration process causing additional costs but that it can be used for many new scientific investigations. In 1956 Tamboulian and Hartmann [8] demonstrated that experiments in the field of vacuum ultra violet spectroscopy can be performed by using synchrotron radiation. In 1971 Rosenbaum, Holmes and Witz [9] performed the first experiments in X-ray scattering.

The synchrotron radiation opened new possibilities in scientific investigations because it has the following unique properties:

1) The wave length ranges continuously from about $10^{-2}$ mm to 0.01 nm.
2) The intensity of the radiation as well as the brightness of the source are very high compared to the values available when one uses conventional sources.
3) The beam is highly collimated.
4) The radiation is almost completely polarized.
5) The radiation consists of short pulses.

A considerable literature on the properties and applications of synchrotron radiation is available [10-22]. There exist also a number of reviews on investigations of biological macromolecules [23-27]. No general review exists, however, on the application of this technique in polymer science. This article shall serve as such a review.

# 2 Properties and Production of Synchrotron Radiation

## 2.1 How Does a Synchrotron Work?

The synchrotron was developed by elementary particle physics in order to accelerate electrons, positrons, protons and other particles. It consists of a ring with a diameter of about a few meters up to more than 100 m in which a vacuum of $10^{-7}$ mbar can be sustained and to which strong electric and magnetic fields can be applied (see Fig. 1). A bunch of electrons or positrons is first accelerated in a linear accelerator to an energy usually lying between 40 MeV and 380 MeV.

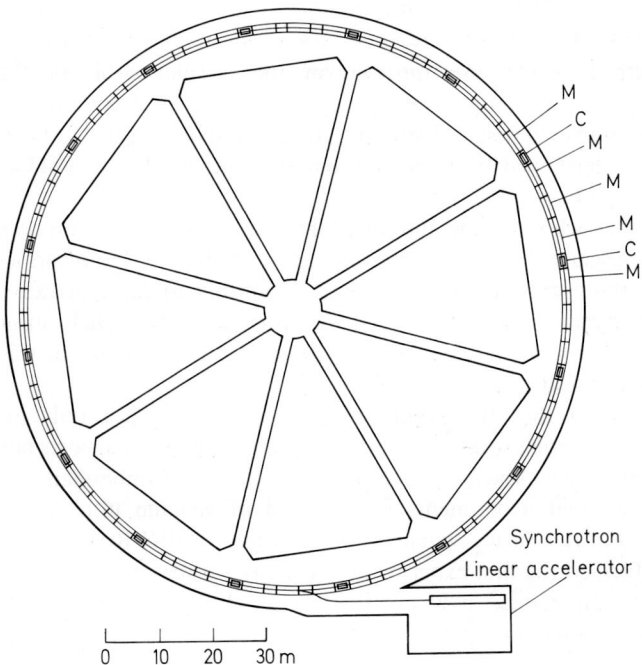

**Fig. 1.** Schematic representation of the arrangement of the 48 magnets M for bending of the beam and the 16 cavities C for acceleration of the particles in the synchrotron DESY at Hamburg

**Fig. 2.** View on magnets in the synchrotron at DESY in Hamburg (by courtesy of DESY)

Afterwards the bunch is injected into the synchrotron.[1] At several positions where the ring is bended strong magnetic fields designated by M in Fig. 1 are applied which force the particles to move on a curved line. In the parts of the ring between the magnetic fields the particles move on straight lines. Here strong electric fields E are applied which accelerate the particles. In the time when they are circling the particles may reach energies up to 7.5 GeV. The electric and magnetic fields are not constant but alternate with high frequency. The frequency is chosen in such a way that the field has its optimal value when it is passed by the bunch of particles. Therefore microwave technique has to be applied and the parts of the ring where the electrons are accelerated have the form of electric cavities. In order to illustrate the dimensions of the instruments applied, Fig. 2 shows a magnet of the synchrotron DESY.

After the particles have reached the energy one wants (for example 3 GeV) they are injected into the storage ring.

## 2.2 The Difference Between a Synchrotron and a Storage Ring

The storage ring, in principle, is built in a similar way as a synchrotron (see Fig. 3). It consists of bended parts where magnetic fields H are applied in order to force the particles to move on a curved line, and of straight parts where electric fields E accelerate the particles. The main difference to the synchrotron lies in the fact that in the storage ring the energy transfered to the particles by the electric field is only as large as it is necessary in order to compensate the losses due to radiation and other reasons[2].

Many bunches from the synchrotron can be injected one after another in the storage ring. After this filling process is finished the particles can be kept circulating in the storage ring with constant energy for a time of many hours. The individual bunches remain seperated during this time. Therefore, at any cross section of the ring, the electric current formed by the particles is not constant but consists of short pulses of duration $\tau_p$ with time intervals $\tau_i$ during which no current is flowing (see Fig. 4). The average current can reach comparatively high values. For example, at the storage ring DORIS in Hamburg values up to 90 mA at energies of 3.7 GeV have been obtained.

While the energy of the particles can be kept constant, the average intensity of the current decreases with time. This is due to collisions with atoms which occur in a small amount even in a very good vacuum and to collisions with the wall of the ring because the focussing is not perfect etc. As an example, Fig. 5 shows the current at DORIS as a function of time as it occurred under experimental conditions.

Of special interest is the time structure of the particle current illustrated in Fig. 4. The duration of the puls $\tau_p$ depends on the length and the velocity of the bunch. At the

---

[1] In some cases, for example at the "photon factory" in Japan the particles are accelerated in the linear accelerator already to the energies of more than 1 GeV and are injected from here directly into the storage ring.

[2] However, in some storage rings like Brookhaven the energy of the particles is also increased after injections before storing at a higher energy. This is called "ramping".

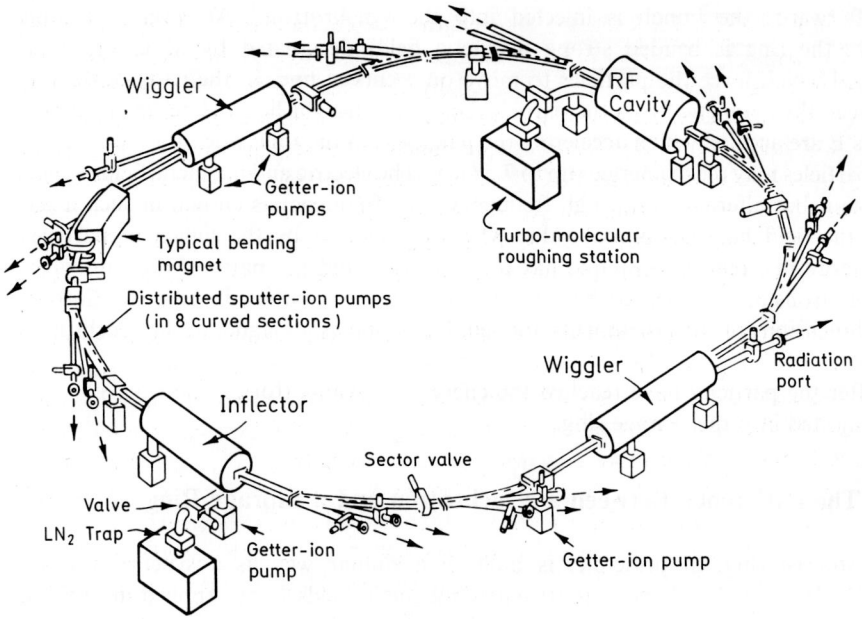

**Fig. 3.** Schematic representation of a storage ring [10]

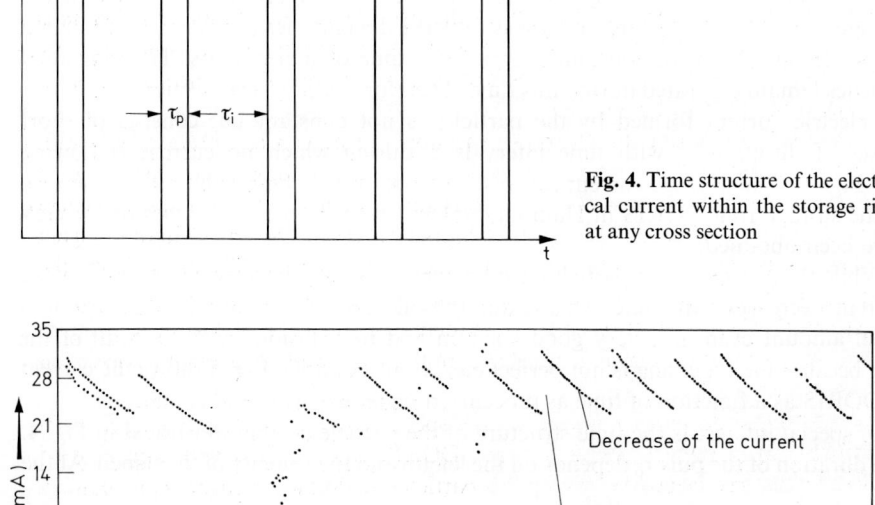

**Fig. 4.** Time structure of the electrical current within the storage ring at any cross section

**Fig. 5.** Diagram of the monitor showing the average current in the storage ring as a function of time. One sees the decrease of the current after each loading process

storage ring DORIS, for example, the length of each bunch is 2–3 cm and $\tau_p$ is about 140 psec.[16]. At SPEAR $\tau_i$ may range between 2.8 and 780 nsec. The time interval between 2 pulses $\tau_i$ is determined by the time which passes between the injection of 2 particles. However, the relation is not so simple that both times are identical. While the revolution time of one bunch in the storage ring is only 1 µs, it takes about 20 ms until another bunch is accelerated in the synchrotron and ready for injection into the storage ring. Therefore, one bunch will perform many revolutions until the next bunch arrives in the storage ring. The times of injection have to be controlled with high accuracy in order to obtain the desired time structure of the current.

The longest value of $\tau_i$ is obtained if only one bunch is introduced into the storage ring. Then, $\tau_i$ is simply given by the revolution time of the bunch which depends on the velocity of the particle (which can be assumed to be that of the light) and the length of the ring. For particles in the storage ring DORIS $\tau_i$ ist about 1 µs as mentioned above. When more and more bunches are introduced, this time can be reduced considerably. At DORIS the shortest time interval is 2 ns. Of course, the shorter this time, the higher the average intensity of the particle current.

It is also important to mention that the storage ring cannot be filled completely with particles. After three quarters of the ring have been filled, the last quarter remains without bunches. Therefore, if, for example, the revolution time of one bunch is 1 µs, the sequence of pulses, shown in Fig. 4, appears during 0.75 µs and a period without pulses of 0.25 µs follows. Afterwards again a period of 0.75 µs of pulses follows and so on.

From this it is seen that there are considerable differences in the particle current between a storage ring and a synchrotron. In the synchrotron only one bunch is circling. Therefore the average intensity of the particle current is much smaller than in the storage ring and there are also no possibilities to change the time structure of the current. In addition, in the synchrotron, the energy of the particle increases with time rather then to be constant.

The improved operating conditions in the storage ring can be obtained only with help of some additional devices which we have not yet mentioned. In order to get a particle current which is constant over a long time it is necessary to focus the beam from time to time. This is done by quadrupole magnets. One quadrupole magnet can focus the beam only in one plane that means either vertically or horizontally. Therefore, always a pair of quadrupole magnets appear. In addition sextupole magnets are applied in order to correct the focussing effect for particles with slightly different energies; the sextupole magnets act in the same way as chromatic correction in optics for visible light.

Some additional devices are applied in the special case of the storage ring DORIS in Hamburg. In difference to other rings, DORIS can be filled with one kind of particles, let us say electrons, while the particles with the opposite sign, namely positrons, are already circling in the ring. Thus, after the second filling process is completed, both particle beams circle simultaneously in opposite direction in different layers (see Fig. 6). At two places designated by W, the beams are bended in vertical direction in order to induce collisions between the electrons and the positrons and to study the generation of new particles. Due to the fact that there are two kinds of particles circling, one needs at each position two bending

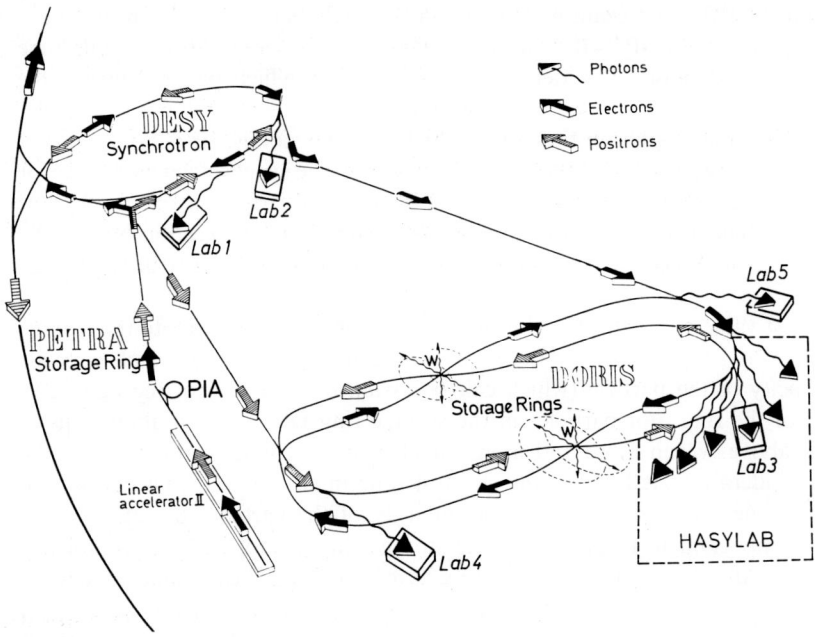

**Fig. 6.** Schematic representation of the linear accelerator, of the synchrotron and the storage ring DORIS at DESY in Hamburg. by courtesy of DESY

magnets instead of one, two quadrupole magnets instead of one and so on. In addition one needs also magnets for the bending in the vertical direction at the positions W.

The possibility to operate the ring simultaneously with electrons and positrons is an advantage mainly for elementary particle physics. For the synchrotron radiation studies it is of no effect despite for the fact that at each position where the beams are bended by the dipole magnet two synchrotron radiation beams are emitted in opposite directions, one from the electrons and one from the positrons. The decrease of intensity due to collisions between electrons and positrons is negligible.

The latest devices applied to storage rings are wigglers and undulators. A wiggler consists of a sequence of dipole magnets in the straight section of the storage ring (Winick et al.[28], Baynham et al.[29]). In the wiggler the electrons or positrons are forced to move on a curved line as indicated in Fig. 7, so that

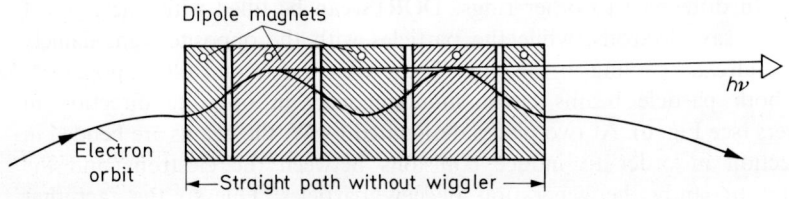

**Fig. 7.** Effect of a wiggler on the particle path

synchrotron radiation is generated also in the straight section of the ring. The simplest wiggler consists of three dipole magnets: one bigger magnet to produce the synchrotron radiation and two others to adjust the flight direction of the particles to that before and after the wiggler. Such a simple wiggler is called single *period wiggler*. If the number of dipole magnets is increased a so called *multipole wiggler* is obtained with a corresponding increase of radiation. If the number of dipole magnets become very large (30 to 100) interference effects in the synchrotron radiation become relevant; in this case the wiggler is called *interference wiggler or undulator* (Hoffmann [30, 31]) Winick [32]). Also *helical wigglers* are considered where, in the field of a superconductic magnet with a bifilar coil, the particles move on a helix. As it is discussed in the next Section, wigglers and undulators are of special importance for the future development of synchrotron radiation experiments. They make it possible to get larger frequencies with larger intensities at considerable lower energy costs.

## 2.3 Theory

According to classical electrodynamics, an accelerated particle with electrical charge emits an electromagnetic wave. In case of an particle oscillating with the frequency ν a wave with frequency ν is emitted. The intensity of the wave depends on the angle θ-between the direction of emission and the vector ñ which lies perpendicular to the direction of acceleration $\vec{v}$ (see Fig. 8a) and is given by

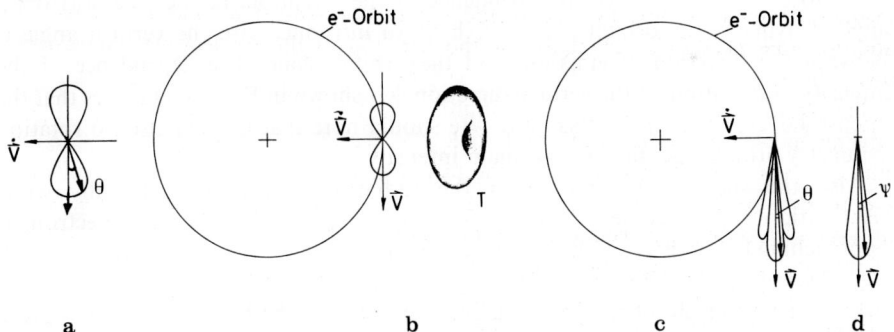

**Fig. 8a-d.** Generation of Synchrotron radiation by a circling electron or positron. **a)** Classical radiation emission by a particle accelerated in direction $\vec{v}$ **b)** corresponding radiation emission by a circling particle in the nonrelativistic case **c)** radiation emission by a circling particle in the relativistic case **d)** out of plane angular dependence of the scattering intensity in case **c)**

$$I = \frac{e^2}{4\pi c^3} \dot{v}^2 \sin^2 \theta . \qquad (2\text{-}1)$$

The left hand side of this equation represents the emitted power per unit spheric angle. The radiation is symmetrical to the orientation of acceleration $\vec{v}$ and therefore the intensity distrubution in all directions of space is obtained from Figure 8a by rotation of the pattern given around $\vec{v}$. The electric vector of the emitted light is polarized in the direction $\vec{v}$. An electrically charged particle moving in a storage ring is all the time accelerated in direction to the center of the ring. This means that, for example, the electron in Fig. 8b is accelerated in a direction given by $\vec{v}$. As long as the velocity v of the electron is low compared to light velocity the considerations made in connection with the oscillating electron can be applied also for the electron circling in an orbit. Therefore electromagnetic radiation is emitted. The dependence of the intensity on the emission angle in the orbit plane is shown in Fig. 8b and corresponds completely to that in Fig. 8a. Again the emission is symmetrical to the direction of acceleration $\vec{v}$ so that the dependence of intensity on all directions of the space is given by the torus like Figure T in Fig. 8b. The frequency of the light emitted is given by the revolution frequency

$$v_0 = \frac{v}{2\pi R} \qquad (2\text{-}2)$$

where v is the velocity of the particle and R is the radius of the orbit.

When the particles in the ring have almost light velocity the situation becomes completely different. Due to relativistic effects calculated from the Lorentz transformation, the torus in Figure 8b is deformed to a narrow cone oriented in the direction of the motion of the particle (see Fig. 8c). In particular the zeros of the radiation patterns which, in the non relativistic case according to Fig. 8b ly at $\theta = 90°$, are now located at a very small angle $\theta$ given by $(1 - v^2/c^2)^{1/2}$ (see Fig. 8c). There is also no longer a rotation symmetry around the direction of acceleration $\vec{v}$. Therefore, in order to describe the direction dependence of the radiation, besides the horizontal angle $\theta$ lying in the orbital plane, we have to introduce also the vertical angle $\psi$ between the direction of emission and the orbital plane. The dependence of the intensity of radiation on the vertical direction $\psi$ is shown in Fig. 8d. One sees that the intensity decreases with increasing $\psi$. We should note that there is also no rotation symmetry around the direction of main intensity $\vec{v}$.

Also the spectral distribution of the radiation is changed markedly when relativistic effects become relevant. Since the source of radiation, the electron, is approaching the observer with almost light velocity, the radiation puls is compressed. The observer in the laboratory systems detects only a short puls of radiation. Hence the Fourier spectrum of the light will contain not only the basic frequency $v_0$ given by Eq. (2-2) but also higher harmonics. In addition the Doppler effect shifts the emitted power from $v_0$ to the higher harmonics. The approximate upper limit of frequency is given by the reciprocal of the duration of the light puls. The higher harmonics are lying so dense that a continous spectrum ist observed.

This is illustrated by the following example. For $R = 12,12$ m it follows from Equ. 2-2 that $v_0 = 2,5 \cdot 10^7$. In the relativistic case, for an energy of 4 GeV the $10^{11}$-th $10^{12}$-th harmonics are emitted with strong intensity, which correspond to wave lengths of about 1 Å. There occur also some changes in the polarization. At low speeds of the particle, the electrical vector of the light is polarized in the orbital plane. At relativistic speeds such a polarization occurs only at the angle $\psi = 0$. At larger values of $\psi$ also a component polarized perpendicular to the orbital plane appears. The dependence of the intensity on the angle $\psi$ and on the wave length $\lambda$

has been calculated by Schwinger [4], (see also the papers of Tomboulian [8] et al. and Godwin [15]). The power $I(\lambda, \psi)$ in cgs-units emitted by a single electron per unit vertical angle $\psi$ and per unit wave length $\lambda$ is given by

$$I(\lambda, \psi) = \frac{27}{32\pi^3} \frac{e^2 c}{R^2} \left(\frac{\lambda_c}{\lambda}\right)^4 \gamma^8 [1 + (\gamma\psi)^2]^2 \left\{ K_{2/3}^2(\xi) + \frac{(\gamma\psi)^2}{1 + (\gamma\psi)^2} K_{1/3}^2(\xi) \right\},$$
(2-3)

with

$$\lambda_c = \frac{4\pi R}{3} \gamma^{-3}$$
(2-4)

$$\xi = \frac{\lambda_c}{2\lambda} [1 + (\gamma\psi)^2]^{3/2}$$
(2-5)

$$\gamma = \frac{E}{mc^2},$$
(2-6)

E is the energy, e the charge and m the mass of the electron. c is the light velocity. $K_{1/3}(\xi)$ and $K_{2/3}(\xi)$ are the modified Bessel functions of II. kind. In Eq. (2-3) the expression within the first bracket corresponds to the intensity of the beam polarized parallel to the orbit plane, that within the second bracket stands for the intensity of the beam polarized perpendicular to the orbit plane.

The dependence of the intensity on the horizontal angle $\theta$ is not included in equation (2-3), the integral value is taken into account. This has the following reason: When a single electron is circling the intensity dependence on $\theta$ changes with time depending on which part of the distribution given in Fig. 8c is just received through the observation slit. When many bunches are present, the intensity dependence on $\theta$ is determined by the distribution of the bunches within the storage ring.

The critical wave length defined by Eq. (2-4) can be used for the characterization of the spectral distribution of the synchrotron radiation. It is a measure (see Fig. 9) for the lower limit of the wave length distribution. By substituting the value for $\gamma$ into Eq. (2-4) one obtains

$$\lambda_c = \frac{4\pi R}{3} \left(\frac{mc^2}{E}\right)^3$$
(2-7)

One sees that $\lambda_c$ decreases with increasing energy of the electrons. If R is measured in meters, E in GeV and $\lambda_c$ in Angstrom this relation yields

$$\lambda_c = 5.59 \frac{R}{E^3}.$$

We shall discuss the dependence of $I(\lambda, \psi)$ on the different parameters by using plots obtained from Eq. (2-3). Fig. 9 shows the dependence of I on the wave length for

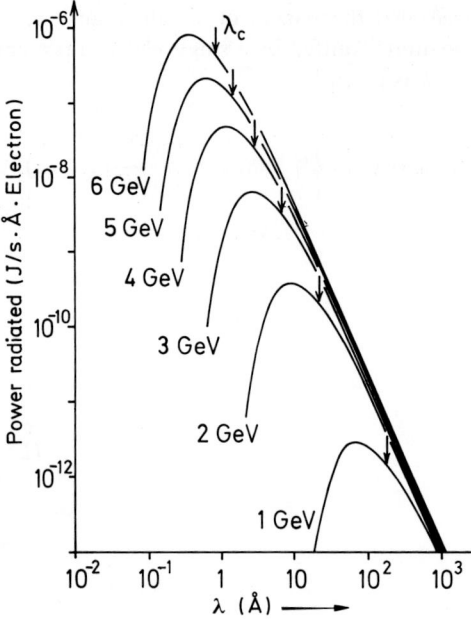

Fig. 9. Spectral distribution of the synchrotron radiation. The parameter gives the energy of the circling particles [15]

monoenergetic electrons. The parameter is the energy of the electrons. When one goes from long wave lengths to shorter wave lengths the intensity $I(\lambda, \psi)$ first increases, then goes through a maximum, and at last decreases again. The larger the energy of the electron, the larger the intensity at each wave length, and the smaller the wave length at which the maximum is reached. The last statement is in agreement with Eq. (2-7). Therefore, in order to reach small wave length, high energies are necessary.

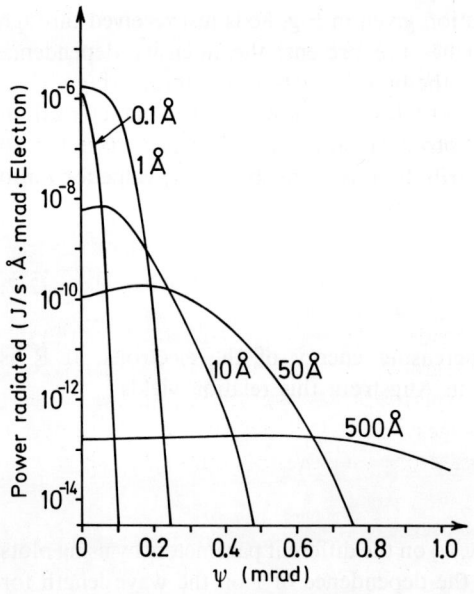

Fig. 10. Angular distribution of the synchrotron radiation intensity for different wave lengths for a storage ring operating at 3.5 GeV. $\psi$ is the elevation angle perpendicular to the orbital plane

The dependence of the intensity on the vertical angle $\psi$ is shown in Fig. 10. The smaller the wave length, the smaller the range of $\psi$ in which radiation is emitted.

As already stated above, the total radiation can be separated into the radiation polarized parallel to the orbital plane with the intensity $I_\parallel$ and radiation which is polarized perpendicular to the orbital plane with the intensity $I_\perp$. $I_\parallel$ decreases monotonically with increasing $\psi$ while $I_\perp$ is 0 at $\psi = 0$ and has its maximum at some finite value of $\psi$. Therefore the curve giving the total intensity I also has a maximum at $\psi \neq 0$.

In Fig. 10, for some curves, due to the logarithmic scale on the ordinate, the first slight increase of intensity I with $\psi$ lies within the accuracy of drawing.

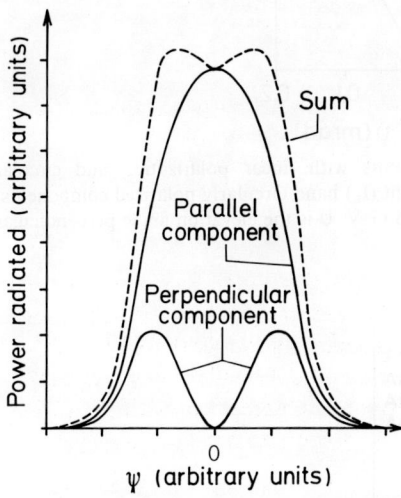

Fig. 11. Angular distribution of intensity components with electrical vector parallel and perpendicular to the plane of the synchrotron [10]

Since the two vector components have a well defined phase relation with respect to each other the light is also elliptically polarized. Figure 12 shows the intensities of the two polarized beams, the degree of linear polarization, and the degree of circular polarization for the storage ring DORIS for three different wave length.

The dimension of the quantity I $(\lambda, \psi)$ considered in Eq. 2-2 is

$$\frac{\text{Joule}}{\text{sec Å mrad}_\psi \text{ electron}},$$

in agreement with the fact that one considers the power scattered by a single electron during a complete revolution integrated over all angles $\theta$ per unit wave length and unit angle $\psi$. In a scattering experiment one is interested in the power emitted per unit electron current in the storage ring and unit angle $\theta$ rather than under the conditions mentioned above. In this case one measures a quantity $\tilde{I}(\lambda, \psi, \theta)$ in the units

$$\frac{\text{Joule}}{\text{sec Å mrad}_\psi \text{ mrad}_\theta \text{ mA}}.$$

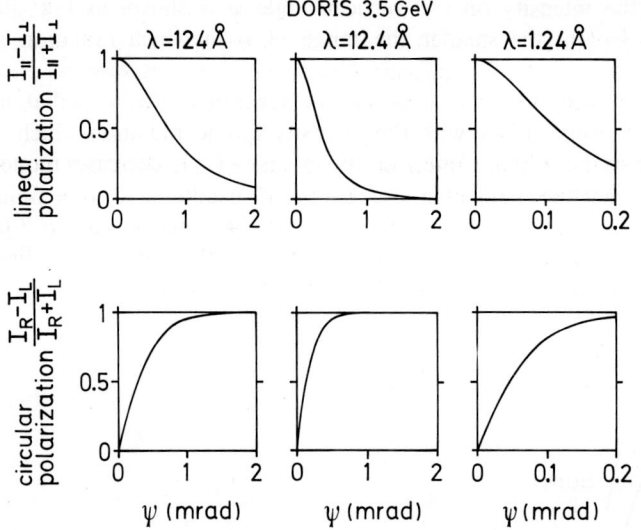

**Fig. 12.** Angular distribution of intensity components with linear polarization and circular polarization (from decomposition into left ($I_L$) and right ($I_R$) hand circularly polarized components) for three photon energies and DORIS operating at 3.5 GeV. $\psi$ is the elevation angle perpendicular to the orbital plan [11]

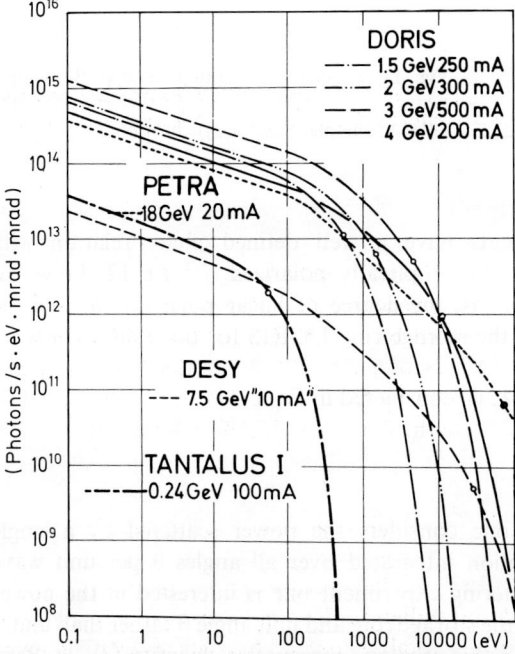

**Fig. 13.** Spectral distribution of intensity in an aperture 1 mrad wide and 1 mrad high centered at a tangential direction. This aperture is well filled at low photon energies while only the part near the orbit is illuminated with hard X-rays. While 1 mrad horizontally is typically accepted by an experiment at a large storage ring, 10 mrad can easily be accepted at a small storage ring like, e.g. TANTALUS [11]

Instead of the dependence on ψ and θ one can also give the power which is absorbed by a detector of a given size in a given distance located at ψ = 0. In addition also the energy absorbed can be measured by the number of photons. One can also consider the energy integrated in all directions. The dimension obtained can be seen from Fig. 9.

Instead of the wavelength λ the photon energy $E_{phot}$, measured in eV, can be inserted. In Fig. 14 for example, one uses the quantity $I(E_{phot}, \psi, \theta)$ which gives the

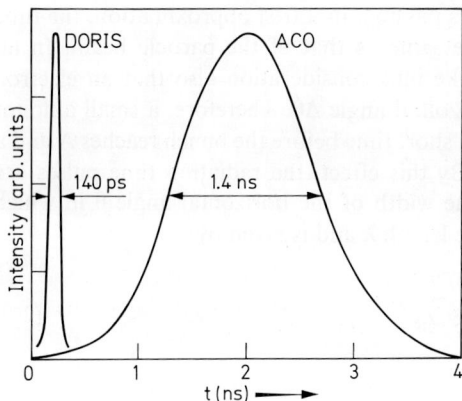

**Fig. 14.** Experimentally determined shape of the light pulses for the storage rings DORIS and for ACO [22]

power radiated in unit angle θ, unit angle ψ per unit photon energy in eV. The wave length λ is obtained from the photon energy $E_{phot}$ by means of the well known relation $\lambda = hcE_{phot}$ which yields

$$\lambda = 1.239 \cdot 10^3/E_{phot} \tag{2-8}$$

if λ is measured in nm and $E_{phot}$ in eV. This relation is illustrated in Fig. 16. Other possible distribution functions given by Green are summarized in the contribution by Winick in the book edited by Winick and Doniach [10].

Of special interest is also the brightness of the radiation source. This quantity is defined as the number of photons emitted by the unit area of the source per unit time, unit steric angle, and unit wave length. Therefore, it can be measured in

$$\frac{\text{photons}}{\text{sec cm}^2 \text{ sterad Å}}$$

(see Fig. 18). In the case of synchrotron radiation the area of the source is the cross section σ of the particle beam. As the total energy of the synchrotron radiation does not depend on σ one can say: The smaller the cross section of the particle beam in the synchrotron, the larger the brightness. In scattering experiments, an image of the source is formed or the detector or, in some cases, on the sample. It is possible to reduce the dimensions of the source by the imaging process. However, if all other conditions including the scattering intensity remain the same, an increased brightness improves the resolution and makes it possible to use smaller samples.

Due to the fact that the particles in the synchrotron do not flow continuously but form bunches as shown in Fig. 4, the synchrotron radiation emitted from a certain point A is not constant with time but consists of short pulses. For many investigations as, for example, X-ray scattering measurements, this time structure is not relevant. For experiments as fluorescence measurements, however, the time structure is of great importance. As, from a certain source A, synchrotron radiation is emitted as long as a bunch of electrons is passing, in a first approximation, the time structure of the synchrotron will be the same as that of the particle beam. In an exact treatment however, one has to take into consideration also that an electron emits the radiation within a certain horizontal angle $\Delta\theta$. Therefore, a small amount of radiation will be emitted from A also a short time before the bunch reaches A and a short time after the bunch has left A. By this effect, the radiation time pulses are broadened. According to Jackson [33], the width of the horizontal angle $\theta$ in which radiation is emitted depends on the wave length $\lambda$ and is given by:

$$\Delta\theta = \frac{1}{\gamma}\left(\frac{\lambda}{\lambda_c}\right)^{1/3} \quad \text{for } \lambda \gg \lambda_c \tag{2-9}$$

and

$$\Delta\theta = \frac{1}{\gamma}\left(\frac{\lambda}{\lambda_c}\right)^{1/2} \quad \text{for } \lambda < \lambda_c. \tag{2-10}$$

$\gamma$ and $\lambda_c$ are defined by Eq. (2-6) and Eq. (2-4) respectively. The experimentally determined shape of the lightpulses for two different storage rings is shown in Fig. 14.

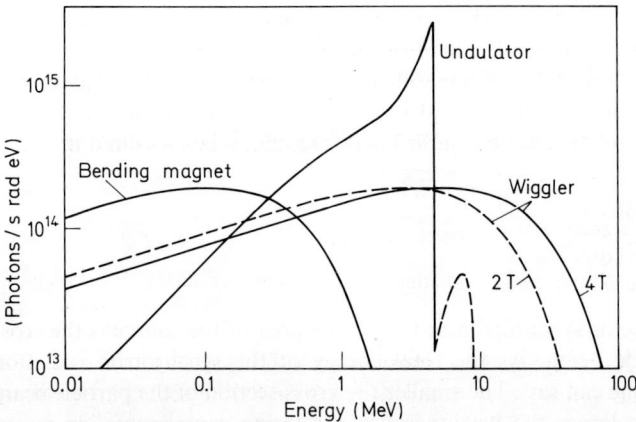

**Fig. 15.** Comparison of the spectral distribution of synchrotron radiation generated by a bending magnet, a wiggler and a undulator [35]

The ordinary intensity distribution as well as other properties of the synchrotron radiation can be changed considerably with help of the wigglers and undulators described in the last section. According to Eq. (2-7), the critical wave length $\lambda_c$ depends on the radius of the circular orbit on which the particles are moving. Within the wigglers in the straight part of the storage ring, this radius is very small so that a small critical wave length is obtained. In undulators [34] where a sufficiently large number of magnetic poles is arranged at constant distances, interference effects between the radiation emitted from each electron at the different poles become predominant; due to this effect the spectrum is quasi monochromatic [34, 35].

The possible effects are demonstrated in Fig. 15 given by Hofmann [35]. The radiation obtained from an ordinary bending magnet with 0.08 T has a critical energy of 0.4 MeV ($= 3 \cdot 10^{-3}$ nm). With a wiggler of 4 T this energy is increased to 20 MeV ($6 \cdot 10^{-5}$ nm). With a long undulator with periodic magnetic fields with a period length of 0.5 cm and magnetic fields of 0.14 T a comparatively sharp line at 4.6 MeV ($= 2.6 \cdot 10^{-4}$ nm) may be obtained.

Wigglers have the further advantage of saving energy. Almost all radiation produced by the wiggler can be used since it is emitted in a small area. On the contrary, ordinary synchrotron radiation is emitted almost everywhere around the storage ring.

## 2.4 Comparison with Other Sources

Figure 16 shows the wave length range covered by synchrotron radiation in comparison with that covered by other sources.

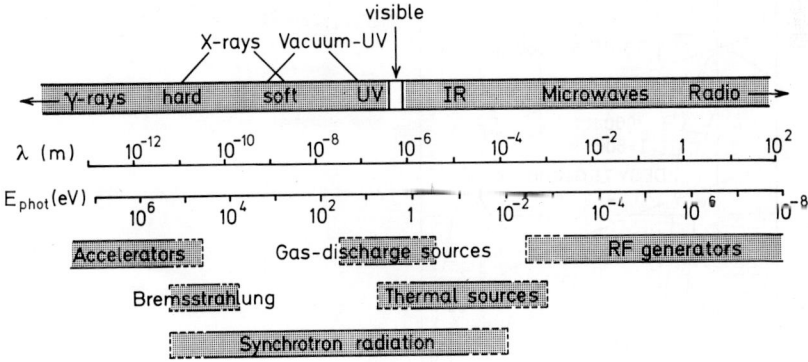

**Fig. 16.** Spectral distribution of synchrotron radiation in comparison with that of other sources

At present time the X-ray range is of most interest for polymer research. One sees that in the conventional sources a gap appears in the region of long X-ray wave length. Here synchrotron radiation is the only source available. This is of importance for the scattering experiments in the anomalous dispersion region which are discussed in the paper by Stuhrmann (this book). At wave length of about 0.1 nm, besides synchrotrons X-ray tubes are available which emit characteristic X-ray lines and the continuous Bremsstrahlspektrum. The intensities emitted by the conventional sources however, are for several orders of magnitude smaller than those from the synchrotron radiation. This is shown schematically in Fig. 17b.

For a quantitative comparison of intensities one has to consider very thoroughly the experimental conditions under which the intensity is used. Among others, the wave length region which is detected, the width of the angle of the emission which can be

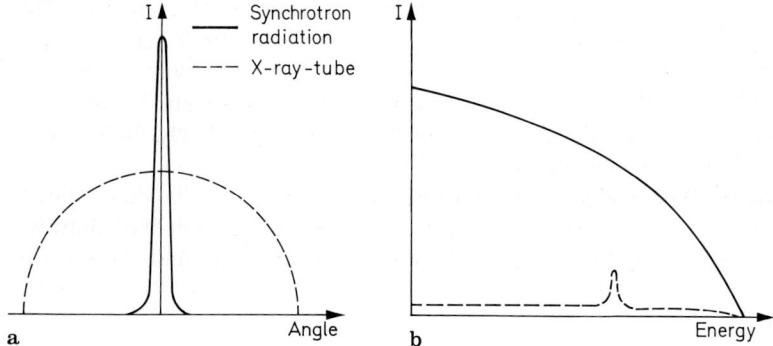

**Fig. 17a and b.** Comparison of the angular distribution (**a**) and spectral distribution (**b**) of synchrotron radiation with that of a X-ray tube [22]

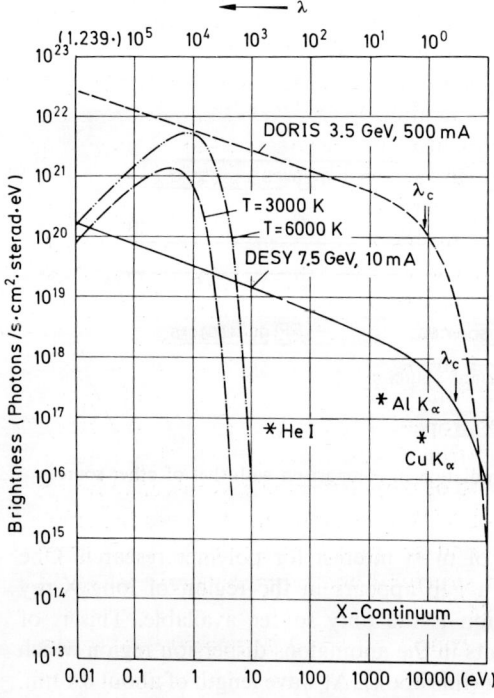

**Fig. 18.** Comparison of the brightness of different synchrotron sourcsec with that of a rotating anode and with that of black body radiation [11]

used in the experiment, and the surface area of the source may be of great influence. For example, the radiation of the X-ray tube is emitted almost in all directions while that of the synchrotron is highly collimated (Fig. 17a). Therefore, if one can use focussing techniques to collimate the intensity emitted by the tube in all directions one obtains larger intensities than in the case that only a small angle can be used. Also the band width is of importance. The characteristic line has a band width $\Delta\lambda/\lambda = \Delta E/E = 5 \cdot 10^{-4}$. For the $CuK_\alpha$ — line we have $\lambda = 0.15$ nm and $E = 8260$ eV. Therefore the band width is $\Delta\lambda = 7,5 \cdot 10^{-4}$ and $\Delta E = 4,13$ eV respectively. If the band width tolerable in the experiment is larger than this value the intensity relation between synchrotron radiation and X-ray tube radiation increases with increasing band width.

Quantitative comparisons were performed by Bonse, Stuhrmann and Kunz [36, 26, 11]. Some representative results are shown in Fig. 18. Here the brightness is considered, as this quantity can be compared best, that means with a least amount of assumption. One sees that the brightness of the synchrotron radiation source is by a factor of more than three magnitudes larger than that of the rotation anode X-ray tube for the $CuK_\alpha$ — line. The brightness of the continuous Bremsstrahlspektrum is still smaller. This property of synchrotron radiation opens the possibility to investigate rapid changes in molecular order and in crystalline orientation. The time structure of the radiation has also been already utilized in X-ray experiments [37, 38].

## 2.5 Present and Future Synchrotron Radiation Sources

In the beginning of the development the synchrotron and storage rings generated only radiation in the region of longer wave lengths starting at about 1 nm. The first succesful measurements with synchrotron radiation were performed in 1963 when R. P. Madden and K. Codling [39] determined the absorption of low energy X rays by gases by using a synchrotron operating at 0.18 GeV at the National Bureau of Standards. Later, in 1974 this synchrotron was converted to a storage ring named SURF II (Synchrotron Ultraviolet Research Facility) which is still operating. The first storage ring dedicated only to the generation of synchrotron radiation is Tantalus I which was completed at the University of Wisconsin in 1968. France has built two storage rings ACO and DCI at Lure in Orsay. Here investigations of polymer dynamics by fluorescence depolymerization are performed.

Sources for radiation in the range of smaller wave lengths down to about 0.1 nm are the synchrotron DESY completed in and the storage ring DORIS completed in, both at Hamburg. The synchrotron radiation of DORIS is presently used by two laboratories, the outstation of EMBL (European Molecular Biology Laboratory) and the HASYLAB (Hamburger Synchrotron Laboratory). At EMBL studies of wide angle and small angle X-ray scattering of biopolymers and synthetic polymers are performed. At HASYLAB facilities for VUV-measurements, X-ray topography, a 2-circle- and a five circle goniometer and so on are available. A small angle scattering beam line for studies of rapid structural and morphological changes in polymer science will be operational in 1984.

Investigations of X-ray scattering of polymers have also been performed at the storage ring VEPP-3 and VEPP-4 in Nowosibirsk. In the United States an extensive

**Table 1.** Comparison of existing synchrotron radiation sources.[22] E: particle energy, R: magnetic radius, I: maximum current, $\lambda_c$: critical wavelength (Eq. 2–4)[a]

| Name | Location | E (GeV) | R (m) | I (mA) | $\lambda_c$ (Å)[a] | Remarks |
|---|---|---|---|---|---|---|
| Group I $\varepsilon_c \lesssim 200$ eV ($\lambda_c \gtrsim 62$ Å) | | | | | | |
| N 100 | Karkov | 0.100 | 0.5 | 25 | 309.96 | Dedicated |
| TANTALUS I | Stoughton, Wisconsin | 0.24 | 0.64 | 200 | 258.30 | Dedicated |
| SURF II | Washington | 0.24 | 0.83 | 30 | 335.10 | Dedicated |
| INS-SOR II | Tokyo | 0.4 | 1.1 | 250 | 95.37 | Dedicated |
| PLAMIA I | Kurtchatov, Moscow | ~0.45 | 1.0 | 100 | ~62 | Dedicated, under construction |
| Group II $\varepsilon_c \cong 200$–2000 eV ($\lambda_c \cong 62$–6 Å) | | | | | | |
| ACO | Orsay | 0.55 | 1.11 | 100 | 37.23 | Dedicated, undulators |
| MAX | Lund | 0.56 | 1.2 | 200 | 41.33 | Dedicated, under construction |
| UVSOR | Okazaki | 0.6 | 2.2 | 500 | 56.87 | Dedicated, under construction |
| VUV | Tsukuba | 0.66 | 2.0 | 100 | 51.88 | Dedicated, under construction |
| VEPP-2M | Novosibirsk | 0.67 | 1.22 | 100 | 22.96 | Partly dedicated |
| NSLS VUV | Brookhaven | 0.7 | 1.9 | 500 | 31.00 | Dedicated, wiggler under construction |
| HESYRL | Hefei, China | 0.8 | 2.7 | 300 | 29.10 | Dedicated, under construction |
| ERNA | Munich, Berlin | 0.43 | 0.2865 | 500 | 20.00 | Dedicated, approved superconducting magnet |
| SILVA | California | 0.8 | ~1.8 | 300 | 19.68 | Proposed |
| BESSY | Berlin | 0.8 | 1.83 | 500 | 19.68 | Dedicated |
| SUPER ACO | Orsay | 0.8 | 1.8 | 500 | 20.00 | Dedicated, under construction |
| IND-I | India | ~0.9 | ~2.0 | — | 17.71 | Proposed |
| ALADDIN | Stoughton, Wisconsin | 1.0 | 2.8 | 500 | 11.59 | Dedicated, wigglers |
| ADONE | Frascati | 1.5 | 5.0 | 60 | 8.26 | Partly dedicated, wiggler |
| Group III $\varepsilon_c \cong 2$–50 keV ($\lambda_c \cong 6$–0.25 Å) | | | | | | |
| DCI | Orsay | 1.8 | 3.82 | 250 | 3.65 | Partly dedicated |
| IPP | Moscow | 2.0 | 5 | 1000 | 3.54 | Proposed |
| SRS | Daresbury | 2.0 | 5.55 | 500 | 3.87 | Dedicated, wigglers |

# Synchrotron Radiation in Polymer Science

| Name | Location | | | | | Notes |
|---|---|---|---|---|---|---|
| VEPP-3 | Novosibirsk | 2.2 | 6.15 | 100 | 2.88 | Partly dedicated, wiggler |
| NSLS X-RAY | Brookhaven | 2.5 | 8.17 | 500 | 2.95 | Dedicated, wiggler |
| Photon factory | Tsukuba | 2.5 | 8.33 | 500 | 3.01 | Dedicated, wiggler |
| PLAMIA II | Kurtchatov, Moscow | 2.5 | | 300 | 1.75 | Dedicated, proposed, wigglers, undulators |
| SPEAR | Stanford | 4 | 12.7 | 100 | 1.12 | Partly dedicated, wigglers, undulators |
| ESRF | | 5 | 22.36 | 500 | 1.00 | Dedicated, proposed, wigglers, undulators |
| DORIS II | Hamburg | 5.7 | 12.12 | 100 | 0.36 | Partly dedicated, wiggler |
| CESR | Ithaca | 8 | 32.5 | 100 | 0.35 | Part-time SR use, wiggler |
| VEPP 4 | Novosibirsk | 7.0 | 16.5 | 10 | 0.27 | Partly dedicated, wiggler |
| Group IV $\varepsilon_c \gtrsim 50$ keV ($\lambda_c \lesssim 0.25$ Å) | | | | | | |
| PEP | Stanford | 18 | 165.5 | 10 | 0.16 | SR laboratory planned |
| PETRA | Hamburg | 20.5 | 192 | 20 | 0.12 | Presently not used as SR source |
| LEP/CERN | Geneva | 120 | 3.410 | 90 | 0.11 | Proposed for high energy physics use |

[a] The critical wavelength in wiggler sections will be considerably smaller than indicated in the table

research programme in synchrotron radiation scattering exists particularly at the storage ring SPEAR at Stanford University. At SPEAR the development of wigglers has been pioneered. Considerable activity on X-ray scattering exists also at the storage ring CHESS of Cornell University. The National Synchrotron Light Source (NSLS) in Brookhaven is expected to go into operation in the near future. In contrast to other storage rings which are also used for high-energy physics experiments, this storage ring has been constructed only for synchrotron radiation experiments.

Further storage rings which have a research programme similar to DORIS are the SRS at Daresbury in Great Britain and the Photon Factory in Japan. For studies in the VUV-region and especially industrial applications like X-ray lithography, the storage ring BESSY at Berlin is used.

Finally we note a proposal for an European Synchrotron Facility (ESRF)[40,41] with a 5 GeV storage ring which can accomodate about 14–15 multipole wigglers and 15 undulators.

Further details about the different synchrotron radiation sources can be obtained from Table 1 and from the literature.[10,11,22]

A comparison of the intensity of a number of sources is made in Fig. 19.

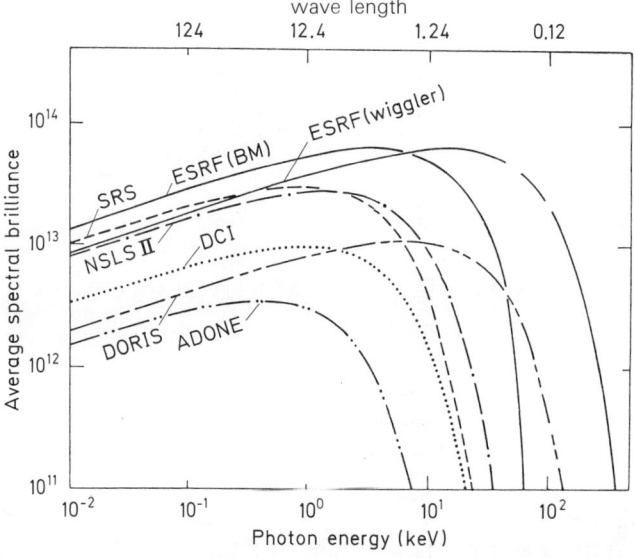

**Fig. 19.** Comparison of the average spectral brilliance of different synchrotron radiation sources [40]

## 3 Instrumentation

### 3.1 Types of Synchrotron Radiation Small-Angle Scattering Cameras

X-ray diffraction experiments can be performed either by the energy-dispersive method or the angular dispersive method. In the first case one uses a polychromatic beam and measures the photon energy distribution of the scattered radiation at a constant scattering angle. In the second case, a monochromatic beam is used and the angular distribution of the scattered radiation is determined.

### 3.1.1 Energy Dispersive Camera

The energy dispersive method is at a first glance persuasive as the figure of merit (FOM) [42] of a given source is more optimally used. The FOM is defined as:

FOM = number of photons/unit time/solid angle/unit wavelength/unit circulating current                                       (3-1)

The schematic design of a small angle scattering camera using the energy dispersive method is shown in Fig. 20. [43] A collimated beam of ~1 mm diameter impinges onto the sample. Sample and detector are separated by an evacuated tube. By using a second collimator in front of a Silicon detector, only radiation scattered under an angle 2θ is recorded.

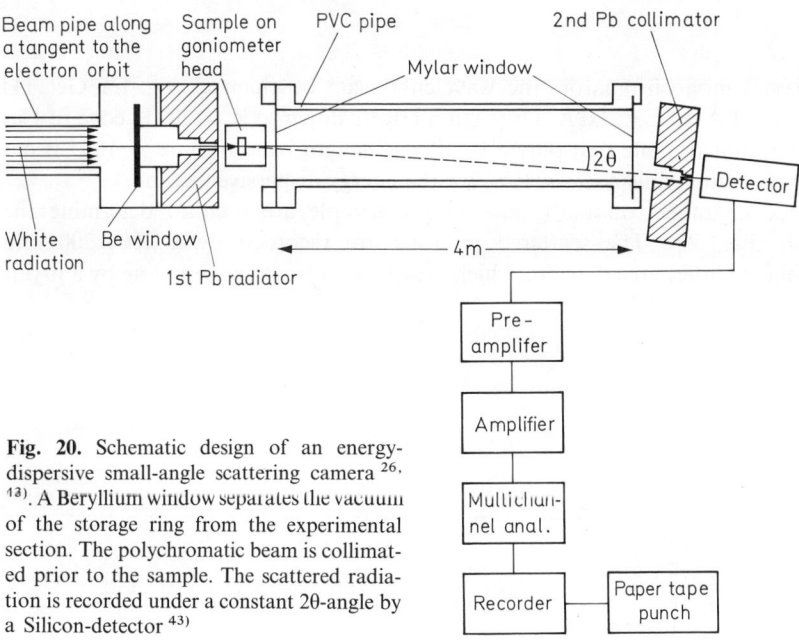

**Fig. 20.** Schematic design of an energy-dispersive small-angle scattering camera [26, 13]. A Beryllium window separates the vacuum of the storage ring from the experimental section. The polychromatic beam is collimated prior to the sample. The scattered radiation is recorded under a constant 2θ-angle by a Silicon-detector [43]

Only few small angle scattering experiments have been performed with this technique [43, 44], however, which is due to the potential deterioration of organic material in the polychromatic beam and the count rate limitation of the detection system. Current applications of this method ly mainly in the area of special environment experiments, e.g. under high pressure and for small structures. [45]

### 3.1.2 Angular Dispersive Cameras

The optics of angular dispersive small angle scattering cameras differ according to the field of application. Thus the *double monochromator* camera is mainly used for anomalous dispersion experiments (Fig. 21) [26, 46] By varying the Bragg angle of two

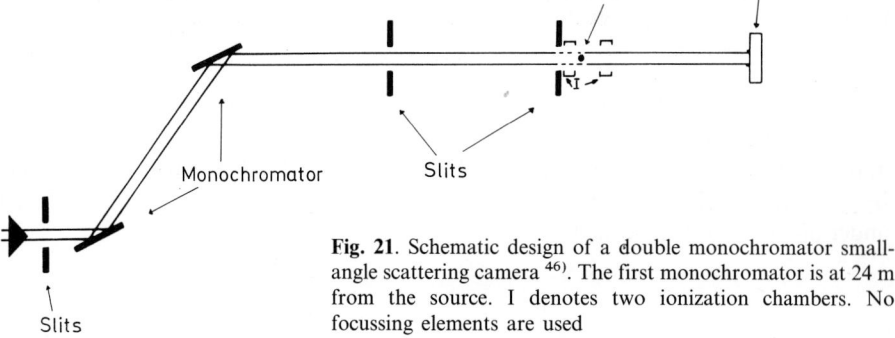

**Fig. 21.** Schematic design of a double monochromator small-angle scattering camera [46]. The first monochromator is at 24 m from the source. I denotes two ionization chambers. No focussing elements are used

perfect crystal monochromators the wavelength can be changed, e.g. for Ge(111) in the range $1.1 \text{ Å} < \lambda < 3.0 \text{ Å}$. Dispersion effects down to $Z = 20$ (K-edge of Ca) can thus be studied. Note that only a small wavelength band of $\Delta\lambda/\lambda \sim 10^{-4}$ is cut out of the spectrum while $\Delta\lambda/\lambda \sim 10^{-1}$ for the energy dispersive method.

Ionization chambers on both sides of the sample are used to determine the absorption (Fig. 22). The scattered photons are recorded by a $200 \times 200$ mm proportional counter, area detector which is separated from the beamline by a mylar window.

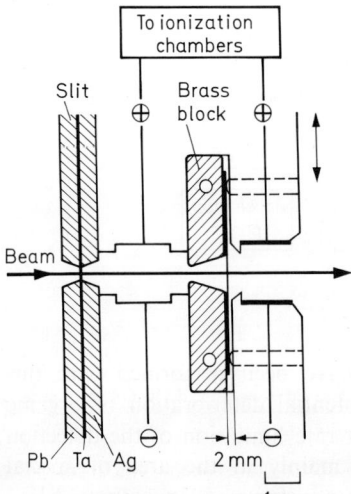

**Fig. 22.** Sample holder with integrated ionization chambers which are used to determine the absorption [47]. A pinhole is used to cut down the beam to the required size

This camera does not use focussing elements. The cross section of the beam is defined by slits at the entrance of the instrument and close to the sample. As the first monochromator is — for architectural reasons — located 24 m from the source point, a considerable fraction of the beam is lost in order to reduce the beam size to the spatial resolution of the detector ($\sim 2 \times 2$ mm). The recent introduction of segmented monochromators [47] suggests to use the second crystal as a focussing element which would increase the intensity at the sample considerably.

*Double focussing, mirror-monochromator* cameras are optimized for maximum flux at the sample. This type of camera is hence mainly used for real time diffraction studies on biological samples [48] and polymers (see Sect. 4). Such a camera is shown in Fig. 23. [26,49] The first optical element could only be placed at 20 m

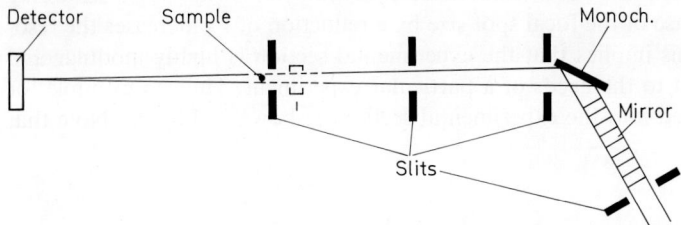

**Fig. 23.** Schematic design of a double focussing mirror-monochromator camera at DORIS [26,49] The middle of the mirror is at 20 m from the source point. A bent, triangular monochromator crystal is used for horizontal focussing and a segmented mirror (quartz) for vertical focussing. The ionization chamber is designated by-I-

from the source point due to the existing shielding of the storage ring. Assuming a vertical divergence of 0.3 mrad and a full width at half height of the source point of $\sim 2$ mm the photon beam will have increased at this distance to $\sim 8$ mm full width at half height. Focussing and demagnification of the source point is therefore necessary in order to obtain a focal spot size in the range of the detector resolution ($\sim 1$ mm for a linear position sensitive detector). The focussing geometry is shown in Fig. 24. [49]

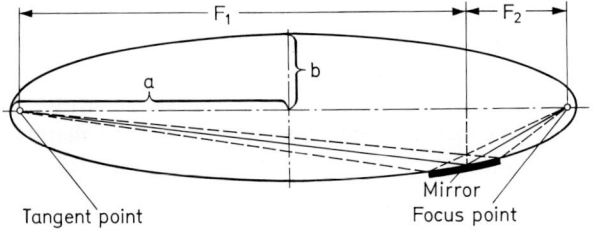

**Fig. 24.** The focussing geometry of a camera shown in Fig. 23. Source point and focus point are the two foci of an ellipse. For vertical focussing the mirror is tangent to the ellipse [49]

A triangular bent Ge(111) monochromator and a segmented quartz mirror are used as optical elements. The fixed wavelength of 1.5 Å which is obtained for a monochromator angle of 26.5° is a compromise between the transmission of Beryllium used as a window material, the efficiency of gasfilled detectors and -especially for biological samples — the transmission of water. [50]

The design of the camera corresponds to a pinhole camera with aperture slits after each optical element and a guard slit front of the detector. A further slit in front of the monochromator defines the beam. The resolution is discussed in 3.3.

The radius of curvature of both mirror and monochromator and hence the focal length — $F_2$ — (Fig. 24) can be changed. This feature which is incorporated in most existing mirror — monochromator cameras allows a flexible adaptation to the characteristics of the source. Changes in the dimensions of the source point which may occur for example when the energy of the storage point is changed, can thus be taken into account. Furthermore, for complimentary wide angle scattering experiments, a decrease of the focal spot size by a reduction of $F_2$ increases the resolution (Sect. 3.3). This implies that the experimental section is highly modular and can easily be adapted to the needs of a particular experiment. Thus an example for a furnace incorporated into the experimental section is shown in Fig. 25. Note that

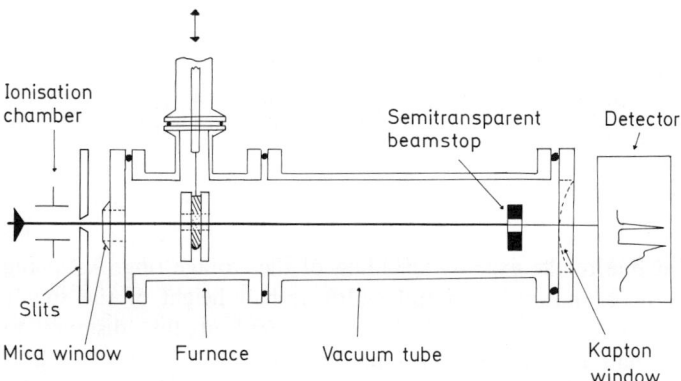

**Fig. 25.** Schematic set-up for a small angle scattering experiment with a furnace incorporated into the beam line. The flux of the monochromatized beam is determined by an ionization chamber. A fraction of the primary beam is transmitted through a semitransparent beamstop and recorded together with the scattered photons with a linear position sensitive detector

the intensity of the primary beam is monitored by a semitransparent beamstop and recorded together with the scattered radiation by a linear position sensitive detector. This has been found to be particularly useful for small angle scattering experiments [51] in case further slits were incorporated between the ionization chamber and the sample. In this case a variation in the dimension of the source point does not change the total flux measured by the ionization chamber but will be noticed in a change of the primary beam intensity measured by a position sensitive detector.

It seems worthwhile to mention a number of modifications to the general design which have been introduced in order to reduce the background, increase the flux at the sample or facilititate the operation of the camera. Thus experimental experience [52,53,54] has shown that the monochromator is the most important source of scattering around the primary beam although a detailed study of its origin (e.g. surface roughness, fluorescence) is lacking. It is therefore preferable to place it as far away as possible from the detector plane, i.e. into the polychromatic beam. A further advantage of this arrangement is that crystals like Germanium support the direct beam from a bending magnet better than the often used quartz mirror. Synchrotron radiation emanating from a wiggler has been found, however, to damage and melt such crystals. [55] Methods to reduce the thermal load on monochromator crystals will therefore have to be developed.

## 3.2 Optical Elements

### 3.2.1 Monochromator

Ge(111) is generally preferred over Si(220) as monochromator crystal due to its higher structure factor. Depending on the nature of the sample one may wish to expand or compress the beam in the horizontal plane by a Fankuchen cut. [49] A compression is particularly useful for the study of small samples (e.g. fibres). The compression is given by:

$$x_e = x_i/A \tag{3-2}$$

where $x_e$ is the diameter of the emerging and $x_i$ of the incoming beam. The asymmetry factor A is defined by:

$$A = \sin(\theta - \alpha)/(\theta + \alpha) \tag{3-3}$$

where $\alpha$ is the angle between the surface normal of the crystal and the reference crystallographic plane (e.g. (111) for Ge). Note that the total intensity will be less than for a symmetric cut ($\alpha = 0$) due to the increase in absorption but that the intensity (photons per unit area) will be $\sqrt{A}$ times higher.

### 3.2.2 Mirror

Both long pieces of float glass which may be bent by shims and a number of short, flat quartz pieces have been used as mirror [49,56]. The latter design seems to be more flexible in order to correct for deviations from a symmetric source point profile. Ameniya et al. [57] have tried to simplify the alignment of the segmented mirror by using a mirror bending device acting on all flat pieces. Only one motor is necessary in order to change the focussing conditions. The results have, however, been disappointing as yet, as a frequent readjustment of the individual mirror supports was necessary [58].

The mirror is also used to suppress the higher harmonics. Thus for a quartz mirror operating in total reflection, the grazing angle is about 3 mrad. Only wavelength longer than given by:

$$\theta_c \text{ (mrad)} = 2.63\lambda \text{ (Å)} \tag{3-4}$$

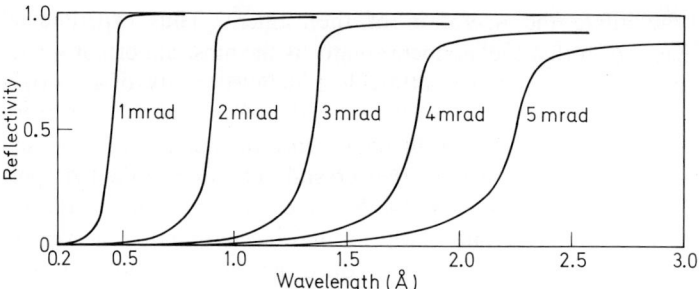

**Fig. 26.** Reflectivity of a quartz mirror as a function of the wavelength at different angles of grazing incidence [49]. An optimum reflectivity at 1.5 Å is obtained for an angle of 3 mrad. The mirror acts as a filter cutting at this angle shorter wavelength

can be reflected (Fig. 26). Mirror materials containing predominantly light elements have been preferred in view of a predicted sharper wavelength cutoff [49]. Recent experimental evidence on platinum coated mirrors suggests, however, that the cutoff is sharper than expected theoretically. [59] This seems to be related to the surface roughness of real mirror materials. In principle the higher angle of total reflection of a Platinum coated mirror would allow to construct a more compact mirror. Whether such a mirror is also a good optical element for a small angle scattering camera in view of the background remains to be seen.

Evidently the design of present cameras is determined by the size of the source point and the large value of $F_1$ which is due to the existing shielding. A simplified optics with a flat or segmented monochromator plus a flat mirror, which is used for the supression of the higher harmonics, could be used in case the optical elements could be placed closer to a smaller source point. This is the case for a camera at VEPP-3 where the distance source point to monochromator is only 3 m. [52,53] A more detailed discussion of this point can be found in Ref. 60.

## 3.3 Resolution

In polymer research the information which can be obtained from small angle and wide angle scattering is often complimentary (see Sect. 4) which suggests to use a small angle scattering camera for both types of experiments. Practical arguments like the

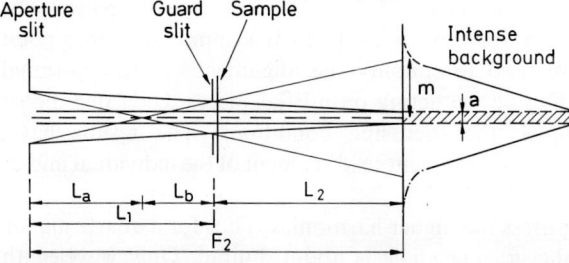

**Fig. 27.** Geometry of a focussing small angle scattering camera after the focussing elements. The resolution is limited by the intense background which is due to scattering from the aperture slit. The guard slit only limits the diffuse scattering without cutting into the primary beam

availability of ancillary equipment or the possibility to perform both types of experiments within the same measuring period should also not be overlooked.

The small angle scattering resolution is determined by the closest approach to the primary beam (Fig. 27). This is limited by the aperture slits which produce a diffuse halo. The guard slits are only used to limit the diffuse scattering without cutting into the primary beam. For a given size of the aperture slits — $S_1$ — the size of the guard slits — $S_2$ — can be calculated according to [61]:

$$S_2 = [(S_1 - a)/(L_1 + L_2)] L_2 + a \qquad (3\text{-}5)$$

and the size of the diffuse halo m by:

$$m = (S_1/2L_a)(L_b + L_2) \qquad (3\text{-}6)$$

where

$$L_b = (L_1 \cdot S_2)(S_1 + S_2) \qquad (3\text{-}7)$$

$$L_a = (S_1/S_2) L_b \qquad (3\text{-}8)$$

The theoretical resolution is given by:

$$L = \frac{\lambda}{2} [\sin(\tan^{-1} m/L_2)/2]^{-1} \qquad (3\text{-}9)$$

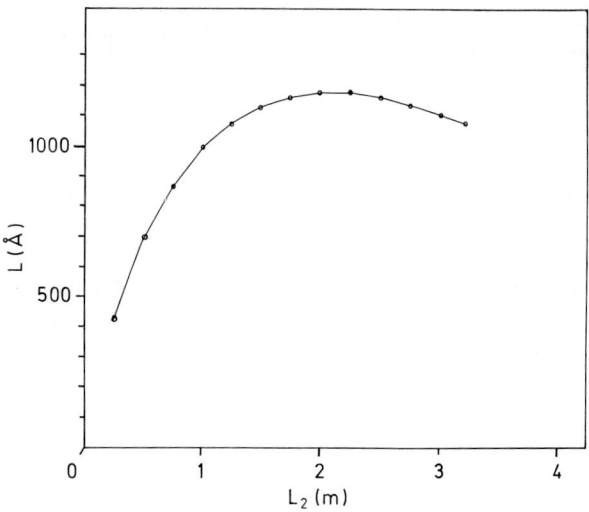

**Fig. 28.** Calculation of the vertical resolution -L- of a small angle scattering camera. $L_2$ corresponds to the distance of the guard slit to the detector plane (see Fig. 27). The other parameters are indicated in the text

Note that the resolution in the vertical plane is higher due to the smaller vertical divergence. A typical calculation of the vertical resolution for a camera shown in Fig. 23 is depicted in Fig. 28. The following parameters were used:

$a_s$ = 2 mm (full width at half height; a is calculated acc. to Eq. 3-15)
$F_1 + F_2$ = 29.5 m (distance source point—detector plane)
$F_1$ = 22 m (distance source point—aperture slit)
$S_1$ = 5 mm (size of aperture slit)
$L_2$ < 4 m (distance guard slit—detector plane)
$\lambda$ = 1,5 Å

The best resolution is obtained for $L_2$ = 2–3 m. A similar resolution is also calculated according to Ref. [49], the decrease of L for smaller $L_2$-values is, however, not obtained as size of the focus is not taken into consideration.

The resolution obtained in practice is smaller as the calculations assume a linear extrapolation of the diffuse halo towards zero which is not true. It can be increased, however, by decreasing $S_1$ (Eq. 3-6) with a corresponding loss in intensity. $F_2$ cannot be increased indefinitively for architectural reasons. Thus for two SAXS-cameras operated by EMBL at DORIS, the resolution lies in the range 1000 to 1500 Å. [49,60] A considerably higher resolution is expected for a small angle scattering camera at a dedicated source like the NSLS at Brookhaven or the projected ESRF which is due to a smaller source point, a smaller vertical divergence and an optimum location of the optical elements.

A higher resolution can also be obtained with a Bonse Hart camera (Fig. 29). [62] Real time experiments are, however, not practical for such a camera as the second

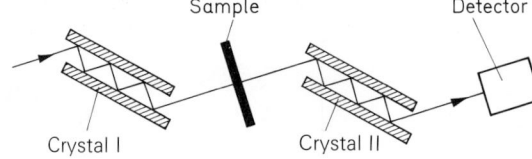

**Fig. 29.** Schematic design of a Bonse-Hart camera. The first double crystal is used for monochromatization. In order to perform a θ-scan, the second double crystal has to be rotated

crystal has to be rotated in order to perform a θ-scan. Preliminary experimental results on biological samples with the Bonse-Hart camera at Doris are reported in Ref. [63].

Consider now the wide angle scattering resolution. According to Rosenbaum and Holmes [25] the width of a diffraction spot is determined by the demagnified source size — a — , the intrinsic width of a reflection and a dispersion term to account for the wavelength spread:

$$d^2 = a^2 + \lambda^2 c^2 L_2^2 + [(\Delta\lambda/\lambda)\,\theta(h)\,L_2]^2 \qquad (3\text{-}10)$$

where c is the width of the reflection in reciprocal space, θ(h) the angular position of reflection h and the other terms are indicated in Fig. 30. Defining:

$$x = 2[\theta(h_2) - \theta(h_1)] \qquad (3\text{-}11)$$

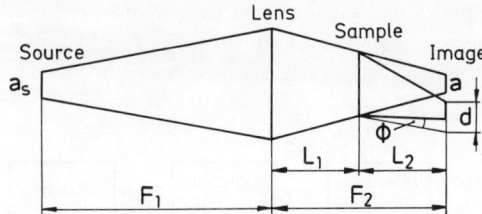

Fig. 30. Schematic design of a focussing optic [25]

as angular separation of two Bragg reflections $h_1$ and $h_2$ one may require that:

$$d < \frac{x}{2} L_2 \qquad (3\text{-}12)$$

is necessary in order to separate two reflections. The optimum condition is then given by:

$$x = 2(a^2/L_2^2 + \phi^2)^{0.5} \qquad (3\text{-}13)$$

where

$$\phi = [\lambda^2 c^2 + (\Delta\lambda/\lambda)^2 \theta^2]^{0.5} \qquad (3\text{-}14)$$

Given a source size $a_s$ it will be advantageous to reduce a as far as possible by making $F_2$ as small as possible:

$$a = (F_2/F_1) a_s \qquad (3\text{-}15)$$

One is limited, however, by the angle of critical reflection as the intensity decreases with increasing angle of total reflection. From this point of view a mirror with heavy atom coating is preferable over a quartz mirror. In view of the use as a small angle scattering camera, low electron density materials have, however, been preferred (Sect. 3.2.2).

## 3.4 Data Aquisition Systems

In this section we will describe two data aquisition systems which are actually used at the storage ring DORIS for experiments on polymers. A detailed description of different data aquisition systems can be found in the article by Hendrix (this Volume).

3.4.1 One Dimensional Position Sensitive Detector

A data aquisition system based on a one dimensional, gas filled detector is shown in Fig. 31. The general philosophy of the system is to be as independant as possible from a central computer as the enormous amount of data demand an on line data analysis during the experiment.

The detector operates according to the delay-line principle. [64, 65] Here a 20 μm thick anode wire and a cathode which consist out of parallel metal strips connected to a delay line are used. The charge creating event induces a signal in the cathode which propagates with a velocity of about 0.2 mm ns$^{-1}$ in both directions of the

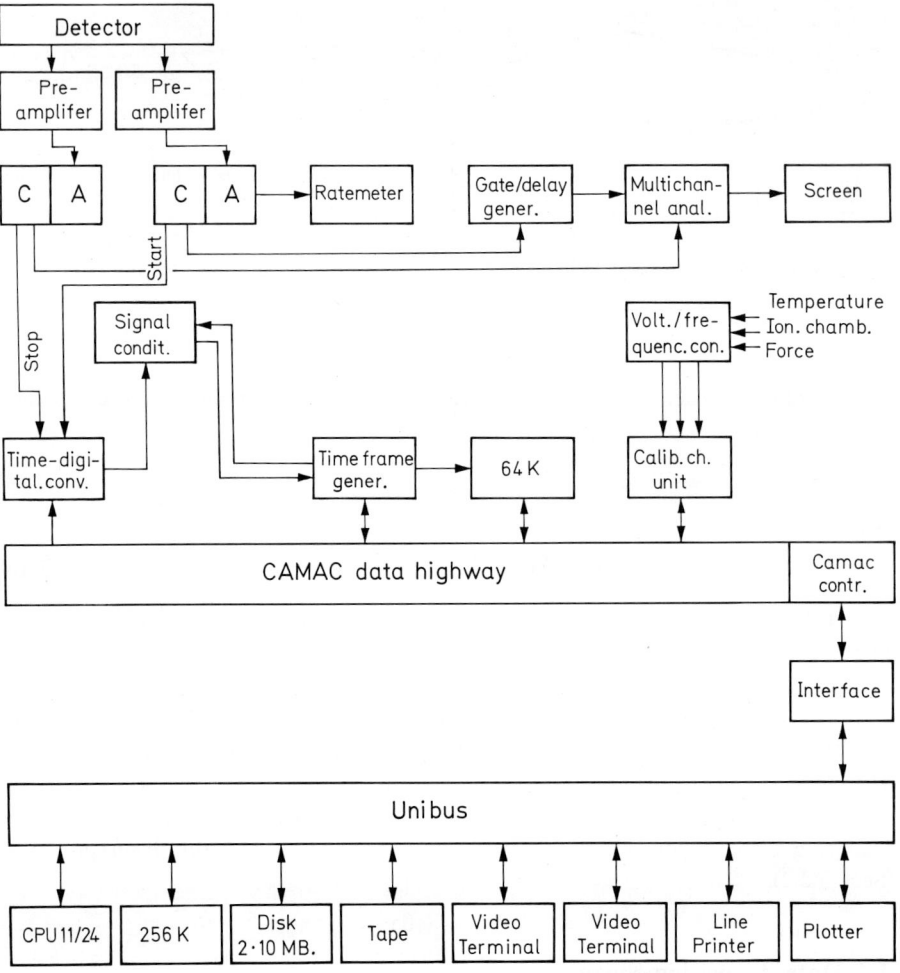

**Fig. 31.** Data aquisition system used for research on polymers at DORIS. The start/stop signals emanating from a linear position sensitive detector [65] is digitized by time-digital converter. The patterns are stored in a 64 kbyte memory. The time information is furnished by a time frame generator. The output of a thermocouple etc. is stored in the calibration channel unit. Data evaluation is possible by a PDP 11/24 computer. (C: Constant Fraction Discriminator; A: Amplifier)

80 mm long cathode. The time difference of the start and stop signals corresponds directly to the position of the charge creating event. For the system shown in Fig. 31, the 8-bit address coming from the time to digital converter is combined with an 8-bit number from the time frame generator. [42] This unit provides the time information to the position information. The 16-bit number is then stored in a 64-kbyte CAMAC memory. Additional information on the photon flux, temperature, force etc. is converted by a voltage/frequency converter and stored in a separate memory. 256 diffraction patterns of 256 channels each can thus be stored. A set of data evaluation programs is available on the level of the PDP11/24 computer [66].

Although the overall count rate limitation with negligible deadtime correction losses is about $10^5$ counts $s^{-1}$, experimental experience shows that it is rather easy to saturate the system, especially for highly localized scattering events (e.g. strong Bragg reflections). This shows the interest in the development of new detection systems.

## 3.4.2 Two Dimensional Position Sensitive Detector

Oriented polymers will no longer show a symmetric scattering pattern. For such systems a two dimensional detector is necessary. A commercial data aquisition system (Westinghouse), based on a Vidicon tube, is shown in Fig. 32. The video signal

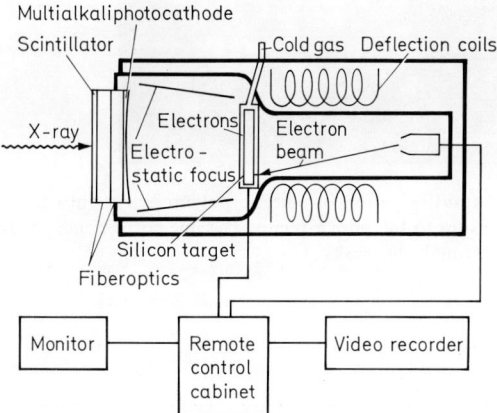

**Fig. 32.** Schematic diagram of the operation of a Vidicon tube used for X-ray diffraction studies (Westinghouse). The videosignal is stored on videotape (UMATIC)

is stored on a videotape. A fibre optics plate of 80 mm diameter, covered by a 30 µm thick layer of $Gd_2O_2S$ or Ag-activated ZnS (Proxitronic), is used to convert the scattered X-rays into photons. Electrons which are subsequently produced by the photocathode are accelerated to a Silicon target which can be cooled in order to increase the signal to noise ratio. The target is read out by an electron beam in the normal TV-mode, i.e. 625 lines in 40 ms.

At present up to 128 video frames can be digitized and further processed with the image analyzing system CA-1. [67]

Typical pictures obtained during the conversion of smectic polypropylene into the α-phase are shown in Fig. 33. [68, 100]

The response of the detector to an incident flux of photons is shown in Fig. 34. [68] Only one line of the video frame was analyzed. This allows a comparison with the linear detector (3.4.1). Evidently the $Gd_2O_2S$-scintillator is appropriate for weak scattering and the ZnS-scintillator for stronger scattering events. No saturation was observed for the latter scintillator up to $2.1 \cdot 10^5$ counts $s^{-1}$ which is a considerable improvement over the linear detector.

The resolution has been determined by comparing the width of reflection spots of a catalase crystal (cubic, a = 156 Å) measured by the Vidicon and measured with a photographic plate. [69] A resolution of about 800 µm (full width at base) has

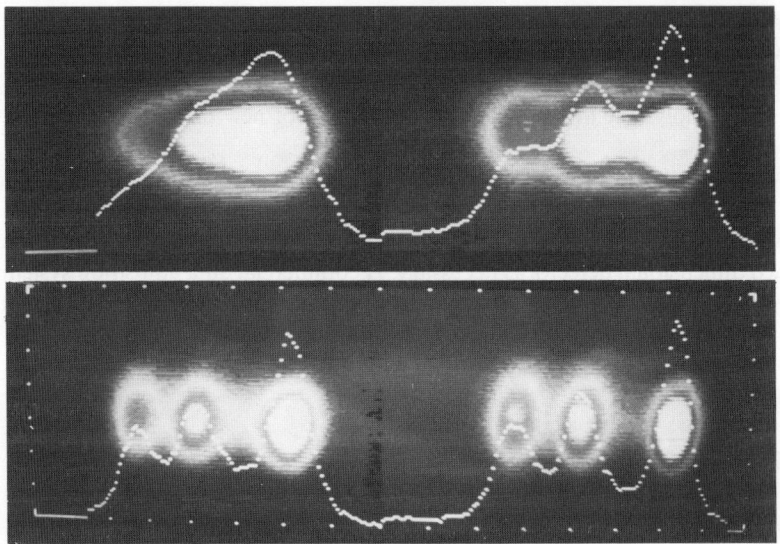

**Fig. 33.** Digitized Videoframes showing the transformation of smectic polypropylene into the α-phase. The temperature was increased in steps up to the pure α-phase. The line corresponds to the intensity contour along one video line and through the peaks

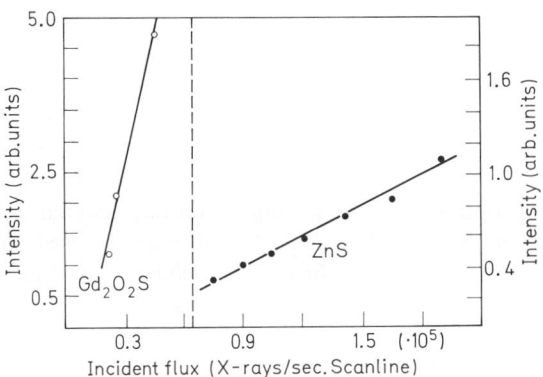

**Fig. 34.** Response of a Vidicon detector to an incident flux of X-rays. The integrated flux of scanline is correlated with the output of the Vidicon for two different scintillators

thus been determined which is comparable to the resolution of the linear detector based on the delay line principle. [65] but considerably better than that of a proportional counter detector described in Ref. [46]. The resolution of the Vidicon is quite sufficient in view of the similar size of the demagnified focal spot (3.1.2).

## 3.5 Ancillary Equipment

In this section we discuss some methods which are used in order to rapidly establish a nonequilibrium situation for real time experiments.

## 3.5.1 Temperature Changes

A thermostated block with a sample holder consisting out of two copper discs which are pressed against a polymer foil is shown in Fig. 35.[70] A 100 µm thick Beryllium sheet is used to ensure thermal contact. For a temperature jump of 7 °C, several minutes were necessary to reach thermal equilibrium.

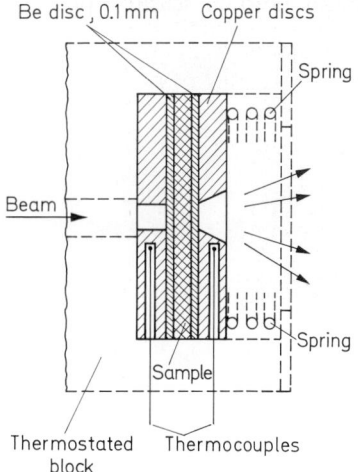

**Fig. 35.** Thermostated block and sample holder for polymer foils [70]. The holder consists of two copper discs which are pressed together with the sample into the thermostated block by a spring. Thermal contact is assured by thin Beryllium discs.

A different design is shown in Fig. 36.[71] Here the sample is fixed within a brass-holder which is pushed by a piston into the heating block which is fixed inside a vacuum chamber (Fig. 25). The heating block consists out of two copper blocks

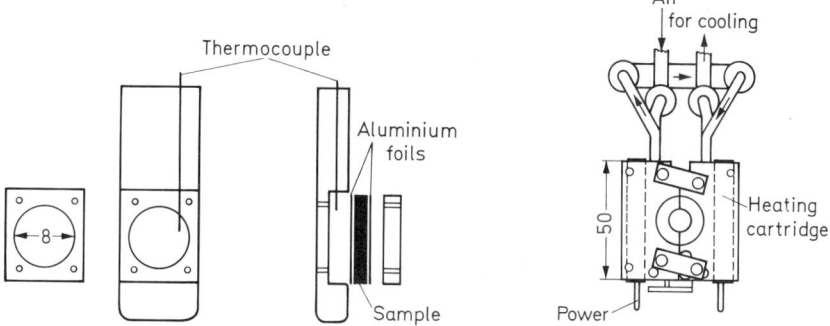

**Fig. 36.** Sample holder for polymer foils made of brass (left picture) [71]. The sample is covered by Aluminium foil to ensure thermal contact. A 0.25 mm thick thermocouple is embedded into the sample. For rapid heating the sample holder is driven by a piston into the preheated heating block (right picture). The separation of the heating block into two parts allows the block to clamp the sample holder thus ensuring thermal contact. Cooling is possible by pressurized air. The beam passes through the 8 mm bore.

each containing a resistance cartridge. Thermal contact is ensured as the holder is clamped by the two copper blocks and by thin Aluminium foils covering the sample. The temperature is determined by a 0.25 mm thick thermocouple embedded into the sample. A temperature of 300 °C is reached without overshooting in about 30 s. Rapid cooling is possible by circulating pressurized air through the copper blocks. Thus Fig. 37 shows the variation of the temperature of a polyethylene sample (Marlex) during a temperature jump from 150 °C to 122 °C. The corresponding wide angle scattering diagramms, as recorded with a linear position sensitive detector and a time resolution of 5 s, are shown in Fig. 38.

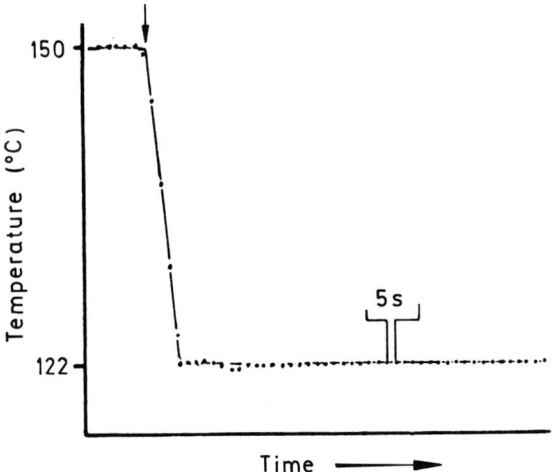

Fig. 37. Change of temperature of the thermocouple upon cooling a polyethylene sample from 150 °C to 122 °C. Start of the cooling is indicated by an arrow.

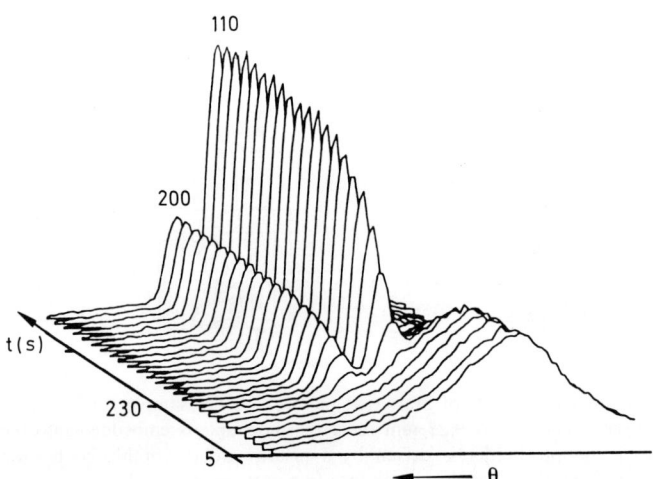

Fig. 38. Selected wide angle scattering diagramms corresponding to Fig. 37. The first diagramms show only scattering due to liquid polyethylene.

Grubb et al.[72] have used a hot air blower shown in Fig. 39. It took 13 s to get within 1 °C of the temperature of the hot air (128 °C). This design cannot be used for air sensitive polymers although one could possibly use inert gas instead of air.

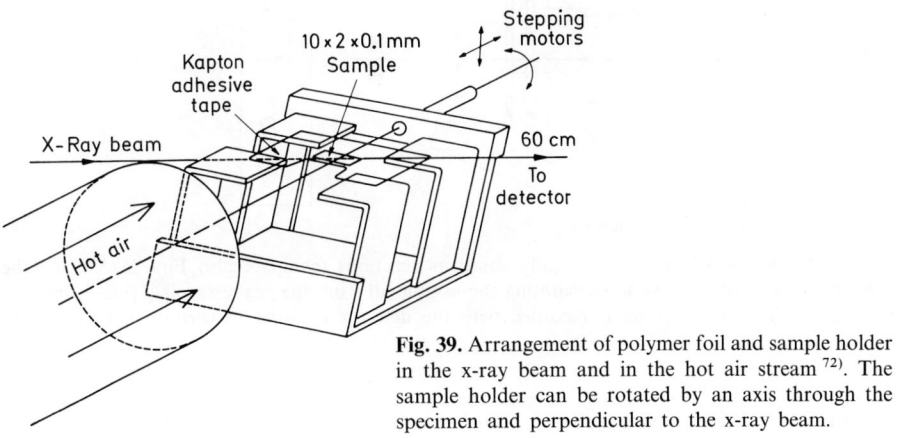

Fig. 39. Arrangement of polymer foil and sample holder in the x-ray beam and in the hot air stream [72]. The sample holder can be rotated by an axis through the specimen and perpendicular to the x-ray beam.

### 3.5.2 Stretching Experiments

A piston driven stretching device has been developed by Koch et al.[73] Stretching occurred in a fraction of a second. The sample can be heated by a stream of hot air. For stretching velocities up to 0.75 mm s$^{-1}$, a stretching device has been developed by Holland-Moritz and Stach [74] which can also be used for real time fourier transformed infrared (FTIR) experiments.[75] Both the force and elongation are transferred to a PDP 11/24 computer by a CAMAC based voltage/frequency converter (Fig. 31).

### 3.5.3 Chemical Reactions

Heterogeneous solid/gas reactions have been studied with the setup shown in Fig. 40.[76,77] The sample is exposed to the reacting gas by opening the tap separating the sample from the gas reservoir. An opening time of about 20 ms can be reached by using a magnetically operated valve which is triggered by the time frame generator of the data aquisition system (Fig. 31).

A principal problem of such experiments is the change in transmission — $\mu$ — as the gas diffuses into the sample. The intensities are then modified by:

$$I = I_0 \, e^{-\mu d} \tag{3-16}$$

where $I_0$ is the intensity of the beam impinging onto the sample and d the thickness of the sample. A correction is possible with the setup shown in Fig. 40. Here, a fraction of the primary beam intensity, which is recorded by the detector, is used for an absorption correction of the diffraction patterns. The change in trans-

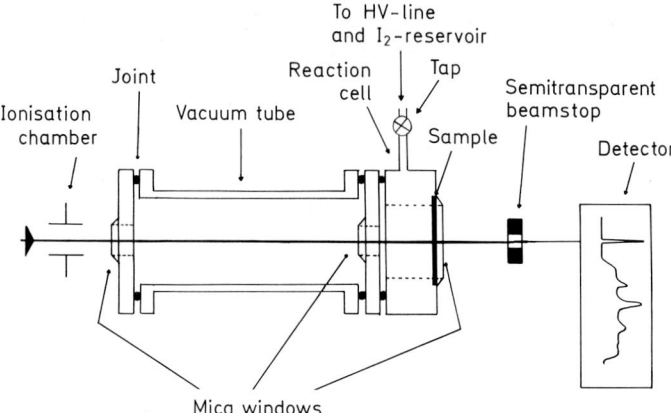

**Fig. 40.** Schematic set-up used to study solid/gas reactions (compare also Fig. 25)[108]. At the onset of the experiment, the tap separating the sample all from the gas-reservoir ($I_2$) is opened. A fraction of the primary beam is recorded with the detector in order to perform a transmission correction

mission during the chemical reaction can also be used to determine an overall degree of reaction [77].

This method does, however, not take the expansion of the sample into account which represents a further systematic error as the beam is smaller than the polymer foil.

## 4 Results and Discussion

### 4.1 Small Angle Scattering During Isothermal Crystallization

The isothermal crystallization process has been investigated by means of many methods. However, several basic questions are still open, among which are the following: What is the crystal thickness at the beginning of the crystallization process and how does this thickness change during the process? Is the difference between the densities of the crystals and the amorphous regions constant or does it change during crystallization?

Answers to these questions can be obtained if one performs small angle scattering measurements during isothermal crystallization. Such measurements have been performed on different materials. The results were observed to depend strongly on the material used. So, for example, with increasing crystallization time, the long period increased with polyethylene, it decreased with polyethylene terephthalate and it stayed constant with poly-β-hydroxybutyrate.

#### 4.1.1 Polyethylene Terephthalate

The first measurements of the small angle scattering during crystallization have been performed on polyethylene terephthalate by Elsner, Zachmann, and Milch [78]. Amorphous films were oriented by stretching at 92 °C and crystallized afterwards at temperatures $T_c$ ranging from 90 °C to 110 °C. During the crystallization the small angle X-ray scattering was monitored by a vidicon-system. From the patterns obtained both the long period and the azimutal half-width of the diffraction maximum was registered.

Fig. 41 shows the long period as a function of the crystallization time for a sample with an initial birefringence $\Delta n_0 = 19 \cdot 10^{-3}$. The parameter designated at each curve is the crystallization temperature. One sees that the long period decreases with increasing crystallization time. Fig. 41 shows the corresponding values of the

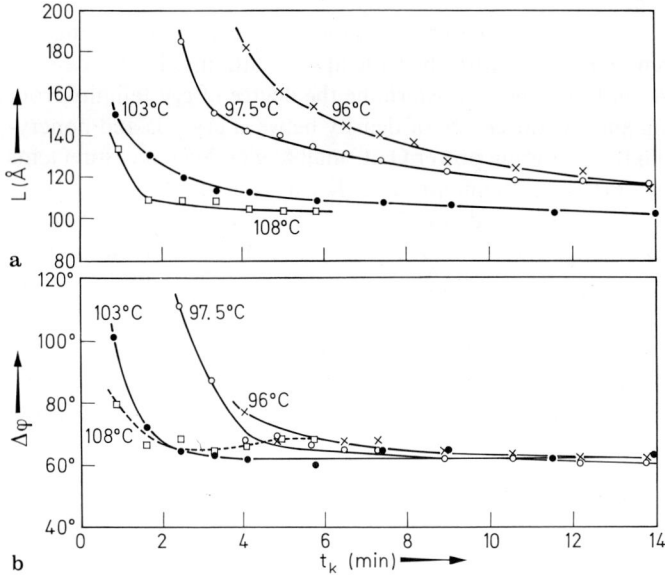

**Fig. 41a and b.** Long period L (a) and azimuthal half width $\Delta \varphi$ (b) as a function of time during isothermal crystallization of oriented polyethyleneterephthalate with the initial birefringence $\Delta n_0 = 19 \cdot 10^{-3}$. The parameter is the temperature of crystallization [78]

azimutal half-width of the diffraction maximum. Generally, the azimutal half-width decreases with increasing crystallization time. On the contrary, the azimutal half-widths of the wide angle reflexions remain constant. From this it is concluded that the orientation of the chains remains constant and only the orientation of the crystal lamella surfaces improves.

A model explaining this behavior is shown in Fig. 42. One has to assume bended lamella which flatten during annealing while the chain orientation does not change. The decrease of the long period is caused probably by this flattening.

Studies were performed also with unoriented polyethylene terephthalate (Elsner, Koch, Bordas and Zachmann [79]). Also with unoriented samples the long period decreased with increasing time. Such a decrease has been observed already with samples quenched to room temperature (Zachmann and Schmidt [80]). The present results show that the decrease is not an artificial effect caused by quenching, but

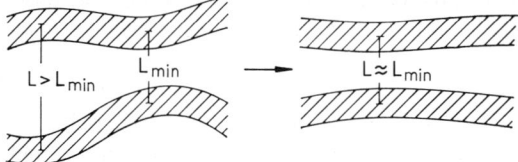

Fig. 42. Schematic representation of the flattening of the crystal lamellae during crystallization

that it can be measured also at the crystallization temperature without interruption of the crystallization process.

Of special interest is the study of the crystallization process by measuring the small angle scattering simultaneously with the wide angle scattering. In this case, as a function of crystallization time, one can determine the degree of crystallinity from the wide angle scattering and the difference of density between crystals and noncrystalline regions, $\Delta\varrho$, from the scattering power Q at small angles. Such measurements have been performed by Prieske, Zachmann, and Koch [81] in the following way: The wide angle scattering is detected by X-ray films. The exposure time is 1 min and the film are changed according every 1 min. Each film has a hole with a diameter of 2 cm in the center. Through this hole the small scattering passes to a linear counter which is standing in an appropriate distance behind the film.

Measurements were performed up to now during isothermal crystallization at 117 °C of an initially amorphous film. Fig. 43 shows the scattering power $Q = \int I(s) s^2 \, ds$ of the small angle scattering as well as the degree of crystallinity obtained from wide angle scattering as a function of crystallization time.

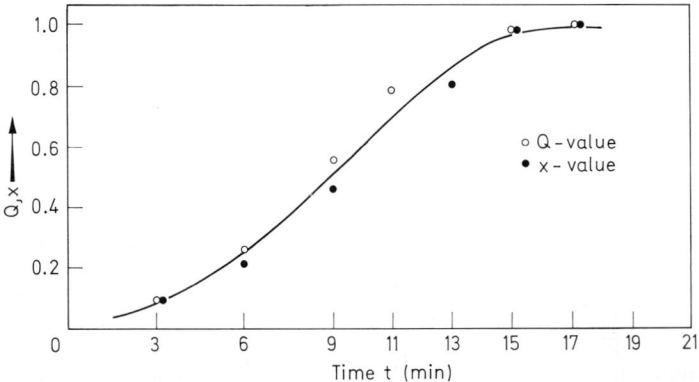

Fig. 43. Scattering power Q and amount of spherulytic crystallized material x as a function of crystallization time during isothermal crystallization of isotropic polyethylene terephthalate at 117 °C

One sees that both curves fall together within the accuracy of the experiment. This indicates that the increase of Q is caused only by an increase of the spherulitically crystallized material and that the difference in the densities of the crystal and the amorphous regions does not change. A model of crystallization by "spinodal decomposition" as it has been used by other authors for highly oriented polyethylene terephthalate and highly oriented polypropylene (Schultz et al. [82] and Fischer [83]) obviously does not apply in this case.

Under the conditions described one can follow the kinetics of crystallization by measuring Q. By this method halftimes of crystallization down to about 2 min can be determined.

### 4.1.2 Polyethylene

The small angle scattering of polyethylene has been investigated by Salazar, Barham, Chivers, and Keller [84] at Daresbury. These authors found that the long period at the beginning of the crystallization process is much smaller than usually found after cooling of the material. It is also smaller than the primary crystal thickness deduced from Raman experiments. By comparison with Raman experiments it is concluded that the primary lamellae are thin, and that the initial stage of thickening involves a single large step (increasing by approximately a factor of 2). After this new thick lamellae continue to thicken logarithmically in the way previously described.

This shows that by means of synchrotron radiation, for the first time the primary lamellar thickening on melt crystallization in polyethylene could be observed. This enabled also the determination of the true crystal thickness $\ell$ as a function of supercooling $\Delta T$. Most satisfyingly it was found to be closely identical to that obtained in solution crystallization were no isothermal thickening takes place. In conclusion, the hitherto existing gap between melt and solution crystallization has been removed and the corresponding $\ell$ versus $\Delta T$ curves brought in coincidence.

### 4.1.3 Poly-β-hydroxybutyrate, PHB

This is a new bacterial thermoplastic and is an ideal model substance for crystallization studies. Burham, Keller, Otun, and Holmes found [85] that the crystals always thicken logarithmically with time when heated above original crystallization temperature, but synchrotron radiation has shown that is does *not* thicken in situ during isothermal crystallization from the melt. This is true even at temperatures where one can observe thickening of lamellae which had previously been crystallized at lower temperatures.

## 4.2 Small Angle Scattering During Annealing Above the Crystallization Temperature

If a polymer is annealed above the temperature of its crystallization one usually observes an increase of the long period which is attributed to a thickening of the crystal lamellae. Does this thickening take place by melting and recrystallization or by diffusion of the chains in chain direction? One can try to find an answer to this question by studying the change of the small angle scattering during the annealing process. Results have been obtained on polyethylene and polyethylene terephthalate.

### 4.2.1 Polyethylene

Investigations were performed on mats of crystal lamellae formed at 70 °C. Grubb [72] observed that, during recrystallization at 113 °C, 125 °C, and 130 °C after very fast heating (within 5 sec) the diffraction maximum obtained at the lower crystallization temperature disappears and a new diffraction maximum corresponding to a longer long-period appears during recrystallization (Fig. 44). This new reflexion develops out of a continuous scattering. From this one has to conclude that the crystals first melt completely and afterwards crystallize again.

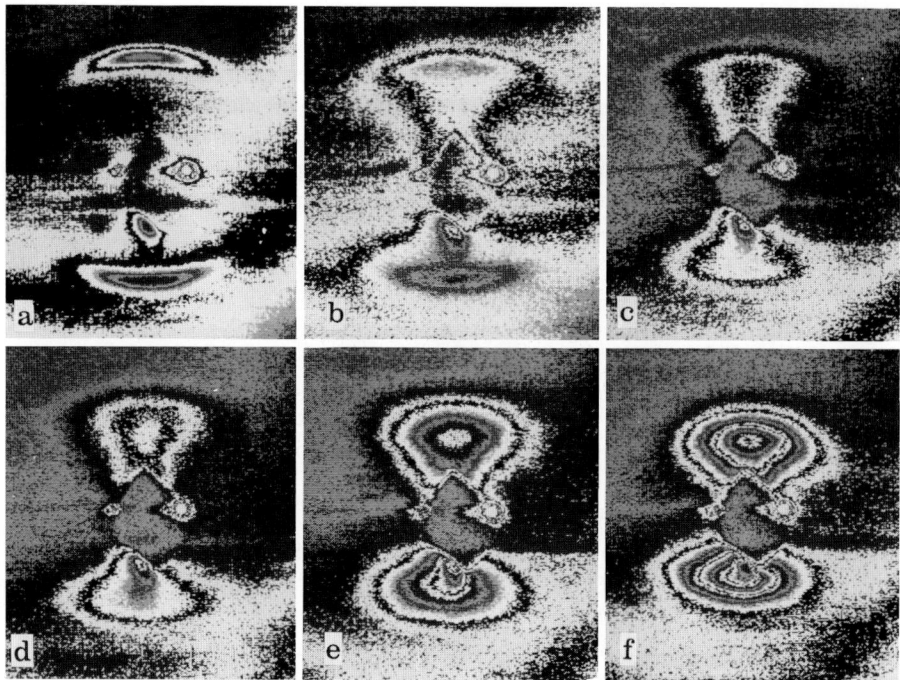

**Fig. 44.** Small-angle x-ray scattering patterns of polyethylene recorded during high temperature annealing with a Vidicon, digitized and contoured averages of five frames [72]. The pictures shown were taken upon heating to 128 °C. The picture — a — shows arcs corresponding to a spacing of 112 Å. After the final temperature has been reached, a new maximum corresponding to a long spacing of 178 Å has formed

Investigations on the same mats were performed also by Spells, Barham, and Keller [86]. In this case the recrystallization took place at 116 °C and 120 °C and the heating time was slower than in the first case (40 °C per min). It was observed that under these conditions the small angle diffraction maximum shifts gradually to smaller angles. The increase of the long period takes place logarithmically with time at 120 °C while a discontinuity occurs at 116 °C. The absolute intensity of the diffraction maximum first increases and remains constant afterwards. There is no indication of a decrease. From these results one can not conclude that no complete melting takes place. However, it is possible that always a small number of lamellae melt and crystallize again with a thickness which is slightly larger than before. The difference in the results obviously is due to the difference in the heating rates.

The reversible changes in the small angle scattering of oriented polyethylene during heating and cooling within the melting region were investigated by Fronk, Heise, Schubach, and Wilke [87]. These authors found that the long period varies reversible between 21.7 nm at 30 °C and 25.2 nm at 100 °C. The change of the long period is attributed to the melting of small crystals and the increase observed during cooling to the recrystallization of these crystals (Fig. 45).

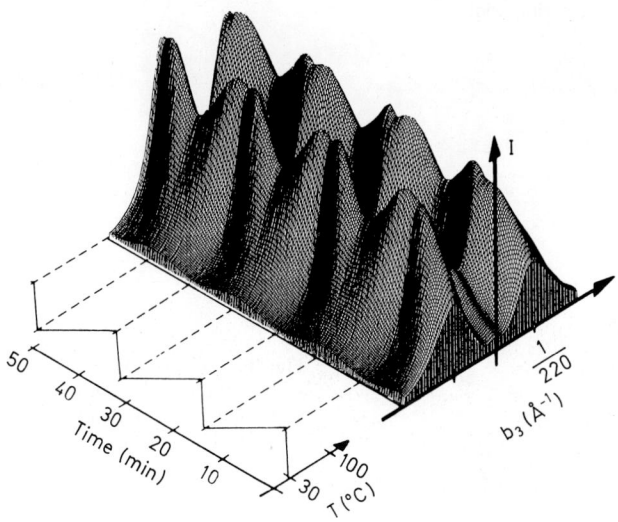

**Fig. 45.** Small-angle x-ray scattering pattern of oriented polyethylene (LDPE Lupolen 1840 D) during four temperature cycles as recorded by a linear position sensitive detector. The detector was oriented parallel to the meridian of the small angle pattern [87]

### 4.2.2 Polyethylene Terephthalate

Investigations on polyethylene terephthalate were performed by Gehrke, Riekel and Zachmann. [88] Amorphous samples were crystallized at 120 °C. Afterwards, the samples were heated up to different temperatures below the melting point and cooled down again. The heating and cooling rate was about 100 °C/min. In a typical experiment the sample was brought to the following temperatures: 120 °C, 230 °C, 120 °C, 240 °C, 120 °C, 245 °C, 120 °C, 250 °C, 120 °C, 250 °C. Up to 240 °C one observes that the scattering power $Q = \int I(s) s^2 ds$ increases simultaneously with heating and decreases simultaneously with cooling. This is due to the different thermal expansion of the crystals and the noncrystalline regions which causes a change of the density difference $\Delta\rho$ between these regions. At 245 °C and 250 °C one observes in addition partial melting and recrystallization. The results for the last two cycles are shown in Fig. 46. Immediately after heating, one observes first in increase of the scattering power Q which is due to the different

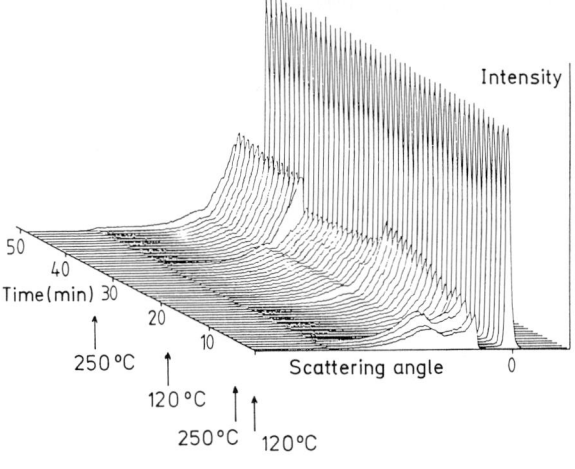

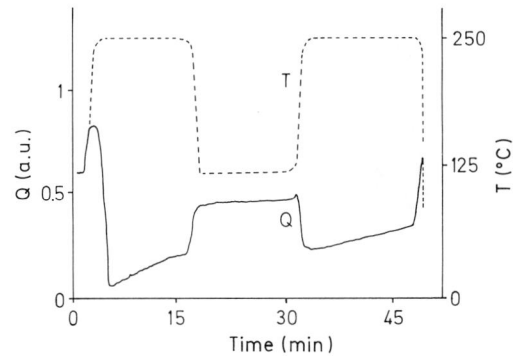

Fig. 46. Change of the small angle scattering during stepwise heating and cooling of unoriented polyethylene terephthalate [88]

thermal expansion of the crystals and the amorphous regions. This increase is followed by a rapid decrease which is due to partial melting. The decrease of scattering power is followed by an increase caused by recrystallization. From this it is seen that one can determine the kinetics of melting and recrystallization by such measurements.

In addition to the scattering power also the change in the small angle peak can be observed. Fig. 47 shows some results. At low temperatures, with increasing temperature, the position of the scattering peak shifts gradually and irreversible to smaller angles. At 245 °C the small angle scattering first disappears completely and appears later again, the peak lying at smaller angles. When the sample is cooled down the new peak remains and the old peak appears again at larger angles.

The results show that above 240 °C the recrystallization occurs after complete melting of the crystal lamellae as in the case of polyethylene with rapid heating. In addition, some lamelae melt and do not recrystallize at the high temperature; they crystallize however after cooling down to 120 °C with the same thickness as they had before heating. As long as the material is heated up only to temperatures below

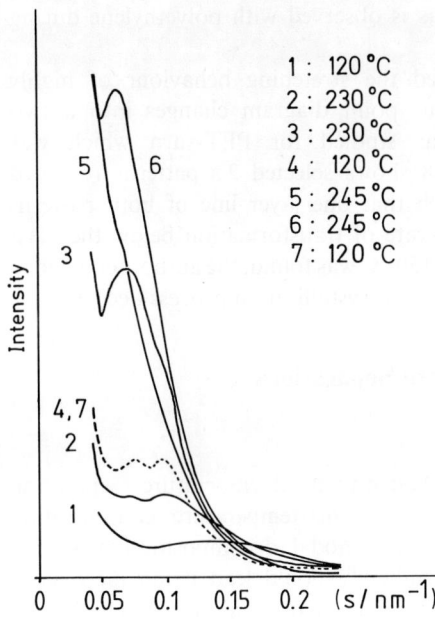

**Fig. 47.** Some small angle scattering curves obtained during heating and cooling of oriented polyethylene terephthalate

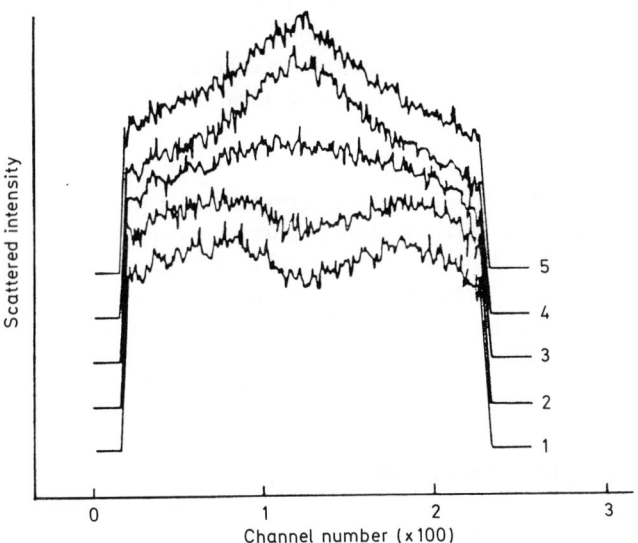

**Fig. 48.** Change in the small angle scattering pattern of polyethyleneterephthalate yarn at 21 °C upon stretching in a fraction of a second [89]. Detector window was perpendicular to fibre axis covered one of the layer lines. The stretch occured between frames 3 and 5. Measuring time per frame: 2 s.

240 °C a gradual shift of the peak occurs as is observed with polyethylene during heating with small heating rates.

Wu, Zachmann and Riekel have studied the stretching behaviour of highly oriented PET.[89] Upon stretching, the four point diagram changes into a two point diagramm. This transformation was studied for PET-yarn which was stretched in a fraction of a second. Fig. 48 shows selected 2 s patterns recorded with a linear PSD which was oriented such that one layer line of both patterns was covered. As no difference between the rate of transformation below the glass temperature (70 °C) at 21 °C and above $T_g$ at 160 °C was found, the authors concluded that an athermal process rather than a melting-recrystallisation process occurs.

## 4.3 Small Angle Scattering During Phase Separations

### 4.3.1 Polymer Blends

Most polymer blends show a miscibility gap. Above a critical temperature a separation into two mixtures takes place. Upon approach to this temperature, concentration fluctuations are expected for the model of a spinodal decomposition which — according to a model by Ornstein and Zernike[90] should lead to an increase in the intensity at smaller angles. As the compositional fluctuations grow in amplitudes as a function of time, a real time SAXS-experiment should show this phenomenon. Endres et al.[91] have studied the melting of mixtures of polyvinylidenefluoride and polyethylacrylate and observed a melting of the crystallites followed by an increase of the intensity at smaller angles, as expected for compositional fluctuations. No evidence for a maximum was obtained, however. A similar observation was made by Russell, Hadziioannou and Warburton for rapidly quenched mixtures of polystyrole and polybutadiene.[92] As the radii of gyration of both components are $\sim 30$ Å and the instrumental resolution $\sim 800$ Å, the authors concluded that the phase formation involves a long range cooperativity of the individual molecules.

### 4.3.2 Block Copolymers

Segmented block copolymers containing only one type of crystallisable unit have been studied by Wegner et al.[93] For segments of oligobutyleneterephthalate or Nylon-12 (crystallizable) and oligo-oxytetramethylen (noncrystallizable) a homogeneous melt without evidence for a concentration fluctuation was observed. Rapid crystallization resulted in a constant long period at an early time of the crystallization process which implies a constant thickness of the crystalline domains. Concentration fluctuations were also not observed by Fischer et al.[94] for Styrole/Butadiene blockcopolymers.

## 4.4 Small Angle Scattering during Crazing

Rothwell, Martinson and Gorman[95] have studied the formation of stress induced crazes in polymethyl methacrylate. The sample was initially stressed and held at a strain of 3.5%. Upon relaxation, SAXS-patterns of 1.5 s were recorded using a linear photodiode detector array. A sequence of selected curves recorded in a plane parallel to the strain direction is shown in Fig. 49. The peak at $h = 0.014$ is probably due to an

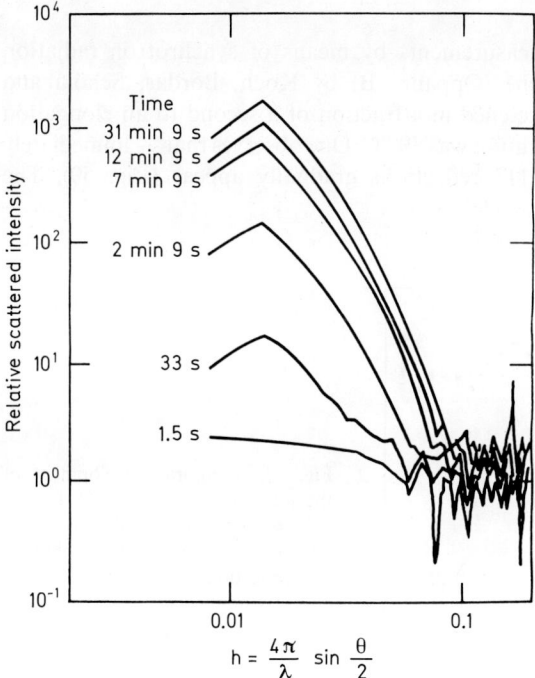

**Fig. 49.** Scattered x-ray intensity curves versus angle parameter h, for a polymethyl methacrylate sample during relaxation at constant strain. Data for the curves were obtained at the indicated times after the initial deformation of 3.5% and in a plane parallel to the strain direction [95]

interference effect between the crazes. From Guinier plots of the SAXS data normal and parallel to the strain direction it was concluded that spherical voids of about 150 Å were formed. Evidence for oriented heterogeneities, which might be planar crazes oriented normal to the strain direction, was obtained too. Whether the voids coalesce into crazes could not be determined with certainty but is a plausible assumption.

## 4.5 Wide Angle Scattering During Crystallization and Solid State Phase Transitions

The kinetics of crystallization is usually measured dilatometrically. However, investigations by wide angle scattering, which became possible by using synchrotron radiation, have the following advantages:
1. If the material crystallizes in different crystal modifications one also gets the information which kind of the different modification is formed.
2. One can also measure the kinetics of the crystallization of oriented materials, which is not possible in a dilatometer because there is no way to apply stress to the samples. In addition, by the X-ray wide angle scattering one obtains information on the degree of orientation.
3. One determines not only an "overall crystallinity" as by density but one obtains also information on changes in crystal imperfections and crystal sizes.

### 4.5.1 Polyisobutylene

The first wide angle scattering measurements by means of synchrotron radiation were performed on polyisobutylene (Oppanol B) by Koch, Bordas, Schöla and Broecker [73]. The samples were stretched in a fraction of a second to an elongation between 300 and 800 %. The temperature was 30 °C. One observes that — immediately after stretching — the 020 and 113 reflections gradually appear (Fig. 50). The

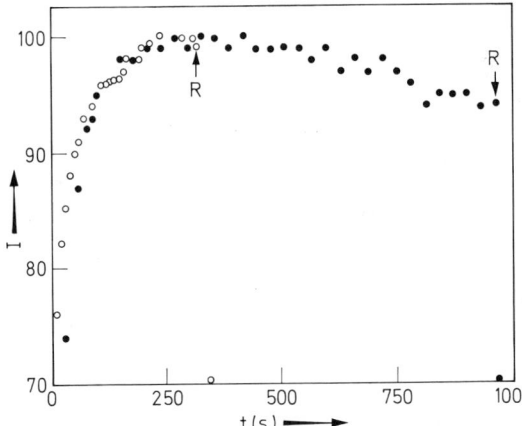

**Fig. 50.** Temporal development of the intensity below the peak for the 020 (●) (temporal resolution: 30 s) and 113 (o) (temporal resolution: 10 s) reflections upon rapid stretching of polyisobutylene [73]. As the fibre is released (R) the intensity drops back to the original value

half time of crystallization is about 40 s. This was interpreted with the athermal formation of rods or fibrils. Upon release of the stress the disappearance of the crystallinity was too quick as to be resolved with a time resolution of 5 s which shows that a different process must operate. Stach et al. [96] showed that such a crystallization occurred also for stretching velocities down to $0.1$ mm s$^{-1}$. No crystallization was observed at smaller stretching velocities. The onset of the crystallization could be correlated with real time fourier infrared spectroscopy (FTIR).

### 4.5.2 Rubber

Caffrey and Bilderback have made a similar study for natural rubber. [97] Using a Vidicon camera they concluded that the amorphous halo disappears while the preferentially oriented powder pattern appears at the same time. Holl et all., [98] have studied the reversibility of this process in more detail. Thus in Fig. 51 the variation of the modulus and the draw ration are compared with selected Vidicon patterns. $\lambda\uparrow$ corresponds to the onset of the crystallization, $\lambda_m$ to the maximum in crystallisation and $\lambda\downarrow$ to the melting of the last crystallites upon relaxation. Note that $\lambda\uparrow$ and $\lambda\downarrow$ occurr, at different draw ratios. This is obviously due to the nucleation process which demands a certain "overdrawing" while the melting occurs at the equilibrium melting temperature.

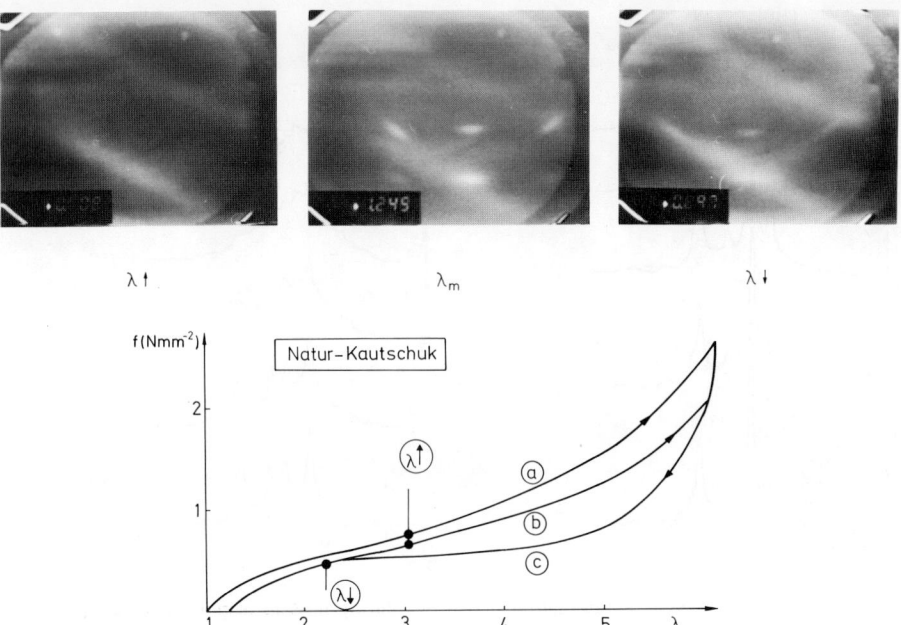

**Fig. 51.** Change of modulus (f) and draw ratio (λ) for natural rubber [97]. The Vidicon pictures show the onset of crystallization during stretching ($\lambda_\uparrow$), the maximum of the crystallization ($\lambda_m$) and the melting of the last crystallites upon relaxation ($\lambda_\downarrow$)

### 4.5.3 Polypropylene

Forgács, Sheromov, Tolochko, Mezentsev, and Pindurin [70] as well as Forgács, Tolochko and Sheromov [99] investigated the isothermal crystallization of unoriented isotactic polypropylene and in specific the transition from the β-phase into the α-phase. Fig. 52a shows the development of the crystal reflexions during the isothermal crystallization. Fig. 52b shows the transition from the β- to the α-modification and the melting of this modification with increasing temperature. Simultaneously with the transition from the β- into the α-modification also an increase in the degree of crystallinity occurs. Fig. 53 shows the increase of degree of crystallinity $x_c$ and the decrease in the fraction of β-modification occuring after heating to 141,3 °C at constant temperature. One sees that the main change occurs very quickly almost within the heating time.

If isotactic polypropylene is quenched from the melt to room temperature a so-called smectic modification is formed which shows only two crystal reflexions. Cabarcos, Bösecke, and Zachmann [100] investigated the kinetics of the transition from this modification into the α-modification. Fig. 54 shows the change of wide angle scattering during isothermal annealing at 90 °C. One sees that the 040-reflexion of the α-modification appears after about 200 sec and continues to increase in intensity afterwards. The time until the constant temperature is reached is about 150 sec. Therefore we can say that most of the process occurs at constant temperature. If the same is done at 130 °C the process occurs so rapidly that it is almost finished as constant temperature is reached.

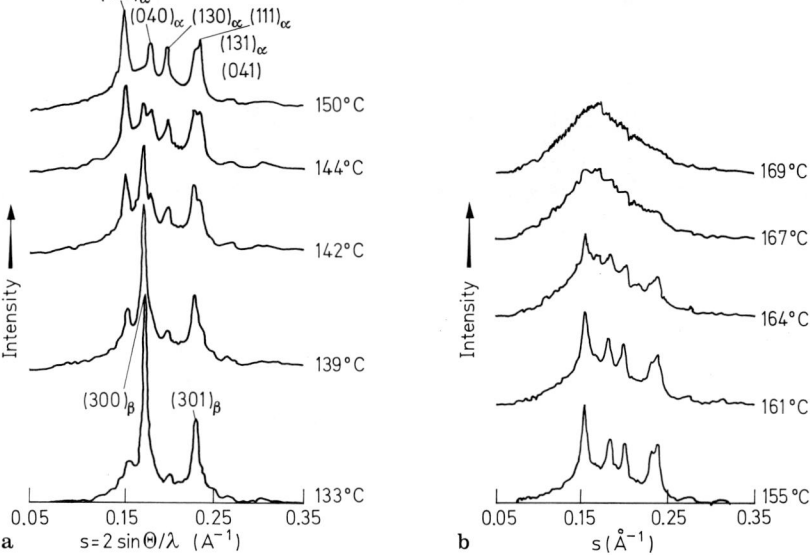

**Fig. 52.** Selected wide angle X-ray scattering patterns during heating of polypropylene indicating the β-α-transition (**a**) and the melting (**b**) [99]

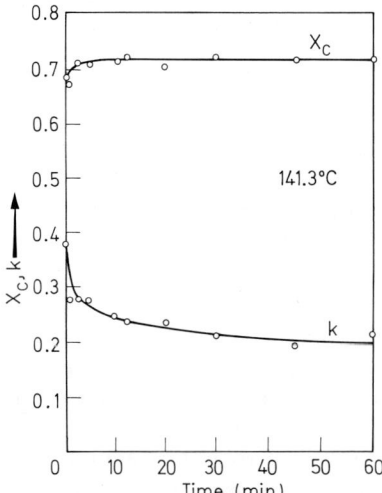

**Fig. 53.** Degree of crystallinity $x_c$ and amount of the β-modification k as a function of time during annealing at 141.3 °C [99]

### 4.5.4 Polyamides

Extensive investigations have been performed also on the crystallization of Polyamide 6. Especially the transformation of the γ-phase into the α-phase was studied. It was shown that the transformation occurs always within a time of less than 20 seconds, that means within the time in which the applied temperature change of the chains in small regions occuring in the liquid crystalline state does not increase considerably the rate of crystallization after cooling.

When the copolyester is heated up or cooled down in the liquid crystalline state fluctuations in the scattering intensity can be observed. Fig. 56 shows the intensity obtained by a position sensitive linear detector and integrated from $2\theta = 12°$ $2\theta = 30°$ as a function of time. Obviously, the time scale of the fluctuations lies in the region of minutes. Measurements at different azimutal angles reveal that the fluctuations are caused by changes in the orientation of the chains. The area of the film from which the scattering arises is about 2 mm × 2 mm.

### 4.5.6 Polyethylene

The stretch induced transformation of initially unoriented orthorhombic polyethylene (Vestolen A) into the monoclinic modification was studied by Heise, Riekel and Stach [104] for stretching velocities down to 0.26 mm s$^{-1}$. The coexistence of both modifications and the formation of two domains of monoclinic material was established by the Vidicon detector.

### 4.5.7 Polyacetylene

The irreversible transformation of cis-polyacetylene into the trans-modification has been studied by Riekel. [76] A sequence of selected diffraction patterns which were obtained upon heating the sample from room temperature to about 350 °C with a heating rate of 11 °C min$^{-1}$ is shown in Fig. 57. At about 150 °C the rate of cis/trans

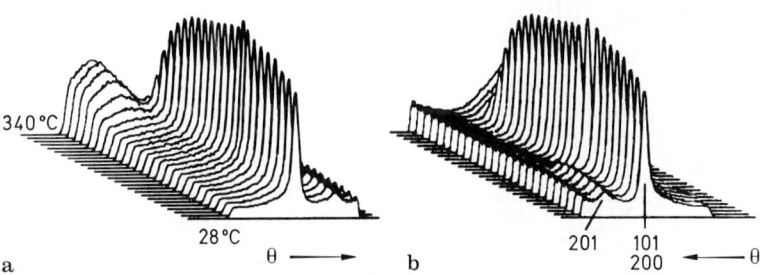

**Fig. 57a and b.** Change of the diffraction pattern of cis-polyacetylene upon heating from room temperature with 11 °C min$^{-1}$ [76]. The cis/trans isomerization is indicated by the disappearance of the 201 reflection at about 150 °C. Above about 290 °C a broad peak indicates decomposition

isomerization reaches a maximum. Above about 290 °C a decomposition starts as evidenced by the appearance of a broad peak. A model of a random formation of trans-chains in the cis-matrix was proposed. For isomerizations at several temperatures an energy of activation of 50 kJ mole$^{-1}$ was determined which agrees rather well with the rate of spin production derived from electron spin resonance measurements. [105]

## 4.6 Melting

By means of synchrotron radiation it is also possible to follow melting under isothermal conditions. Such experiments were performed on Polyethyleneterephthalate. [101] It was shown that with oriented samples melting occured more slowly and at higher temperatures than with unoriented samples. This is in agreement with the explanation [107] of superheating effects by means of entropy considerations.

## 4.7 Wide Angle Scattering During Chemical Reactions

The reaction of cis-polyacetylene with Iodine vapour has been studied by Riekel. [77] A bulk reaction, as evidenced by the appearance of a characteristic 8 Å reflection, was observed immediately after opening the tap of the $I_2$-reservoir (Fig. 58).

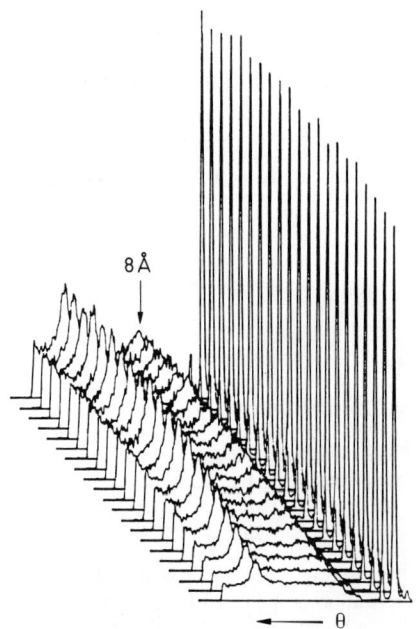

Fig. 58. Change of the diffraction pattern of cis-polyacetylene upon chemical reaction with $I_2$-vapour at room temperature [77, 108]. An analysis of the temporal development of the new reflection (d ~ 8 Å) shows that it appears immediately after opening the tap

The kinetics of the reaction appears not be diffusion controlled. A halftime of reaction $\tau_{1/2}$ of ~340 s was derived while a further much slower process with $\tau_{1/2}$ ~ 240 min was determined by real time neutron diffractometry. [108]

# 5 References

1. Liénard, A., L'Eclairage Elect. *16*, 5 (1898)
2. Schott, G. A., Electromagnetic Radiation, Cambridge University Press, London, Chapters 7 and 8 (1912)
3. Iwanenko, D., and Pomeranchuk, I., Phys. Rev. *65*, 343 (1944)
4. Schwinger, J., Phys. Rev. *70*, 798 (1946), Phys. Rev. *75*, 1912–25 (1949); Proc. Natl. Acad. Sci. USA *40*, 132 (1954)
5. Blewett, J. P., Phys. Rev. *69*, 87 (1946)
6. Baldwin, G. C., Physics Today *28*, 9 (1975)
7. Elder, F. R., Gurewitsch, A. M., Langmuir, R. V., and Pollock, H. C., Phys. Rev. *71*, 829–30 (1947) and J. Appl. Phys. *18*, 810 (1947); Elder, F. R., Langmuir, R. V., and Pollock, H. C., Phys. Rev. *74*, 52 (1948)
8. Tomboulian, D. H., Hartmann, P. L., Phys. Rev. *102*, 1423 (1956)
9. Rosenbaum, G., Holmes, K. C., Witz, J., Nature (London) *230*, 434 (1971)
10. Winick, H., and Doniach, S., (Editors), "Synchrotron Radiation Research", Plenum Press, New York and London (1980)
11. Kunz, C., (Editor), "Synchrotron Radiation in Topics in Current Physics", Springer Verlag, Berlin, Heidelberg, New York (1979)
12. Hodgson, K. O., Winick, H., Chu, G., (Eds.), "Synchrotron Radiation Research", and the Stanford Synchrotron Radiation Project, SSRP Report Nr. 76/100, August 1976
13. Mills, D. M., Physics Today, April 1984, p. 22
14. Madden, R. P., Synchrotron Radiation and Applications in X-Ray Spectroscopy (Ed.); Azaroff, L. V., McGraw Hill Book Company, New York, 1974, p. 338
15. Godwin, P. R., Synchrotron Radiation as a Light Source in "Springer Tracts in Modern Physics", *51* (Ed.) Höhler, G., 1969
16. Rowe, M., and Weaver, J. H., Scientific American, June 1977, p. 32
17. Perlmann, M. L., Rowe, E. M., and Watson, R. E., Physics Today, July 1974, p. 30
18. Munro, I. H., Sabersky, A. P., "Synchrotron Radiation as a Modulated Source for Fluorescence Lifetime Measurements and for Time Resolved Spectroscopy" in Ref. 10, p. 323
19. Brown, F. C., Physikalische Blätter *34*, 619 (1978)
20. Bienenstock, A., Winick, H., Physics Today, June 1983, p. 48
21. Physics Today, May 1981
22. Handbook on Synchrotron Radiation, Vol. 1a, 1b; Ed. E. E. Koch; General Ed. D. E. Eastman, Y. Farge, North-Holland, Amsterdam (1983)
23. Stuhrmann, H. B., Quarterly Reviews of Biophysics, *11*, 71 (1978)
24. Bordas, J., Mandelkow, E., "Time Resolved X-Ray Scattering from Solution using Synchrotron Radiation; Sha'afi, R. I., Fernandez, S. M., (Eds.), in "Fast Methods in Physical Biochemistry and Cell Biology", Elsevier Science Publisher (1983)
25. Rosenbaum, G., and Holmes, K. C., in Ref. 10, p. 533
26. The Use of Synchrotron Radiation in Biology, Stuhrmann, H., (Ed.), Academic Press, London (1982)
27. Stuhrmann, H. B., Ref. 10, p. 513
28. Winick, H., Helm, R. H., Nuclear Instruments and Methods *152*, 9 (1978)
29. Baynham, D. E., Clee, P. T. M., Thompson, O. J., Nucl. Instr. and Meth. *152*, 32 (1978)
30. Hofmann, A., Nucl. Instr. and Meth. *152*, 17 (1978)
31. Hofmann, A., in "Research with Synchrotron Radiation", Abstracts of Nordic Symposium, Gysinge, June 13–15 (1980)
32. Winick, H., Brown, G., Halbach, K., Harris, J., Physics Today, May 1981, p. 50
33. Jackson, J. D., in "Classical Electrodynamics", John Wiley and Sons, New York, 1975
34. Alferov, D. F., Bashmakov, Yu. A., Bessonov, E. G., Sov. Phys. Tech. Phys. *18*, 1336–1339 (1974)
35. Hofmann, A., Phys. Rep. *64*, 253 (1980)
36. Bonse, U., X-ray Sources in "Characterization of Crystal Defects by X-ray Methods", (Eds.), Tanner, B. K., Bowen, D. K., Plenum Press, New York, London, p. 298 (1980)
37. Larson, B. C., White, C. W., Noggle, T. S., Mills, D. M. Phys. Rev. Lett. *48*, 337 (1982)

38. Pruss, D., Huber, G., Danielmeyer, H. G., Bartunik, H. D., (1981) Proceed. XII, Int. Congr., Crystall., Ottawa, C-51
39. Madden, R. P., Codling, K., Phys. Rev. Lett. *10*, 516 (1963)
40. European Synchronotron Radiation Facility, Supplement I, "The Scientific Case", Y. Farge, P. J. Dake, (Eds.), European Science Foundation, Strasbourg (1979)
41. Buras, B., "The European Synchrotron Radiation Projekt", ESRP-PG-3/83, European Science Foundation, Strasbourg (1983)
42. see for example: Bordas, J., Koch, M. H. J., Clout, P. N., Dorrington, E., Boulin, C., and Gabriel, A., J. Phys. *E13*, 938 (1980)
43. Bordas, J., Randall, J. T., J. Appl. Cryst. *11*, 434 (1978)
44. Bordas, J., Munro, I. H., Glazer, A. M., Nature *262*, 541 (1976)
45. For a review see for example: Buras, B., Nucl. Instr. Meth. *208*, 563 (1983)
46. Stuhrmann, H. B., Gabriel, A., J. Appl. Cryst. *16*, 563 (1983)
47. Sparks, C. J., Ice, G. E., Wong, J., Batterman, B. W., Nucl. Instr. Meth. *194*, 73 (1982)
48. for a review see for example: Buras, B., Fourme, K., Koch, M. H. J., in "Handbook on Synchrotron Radiation", Vol. 1b, p. 1015, E. E. Koch (Ed.), North-Holland (1983)
49. Hendrix, J., Koch, M. H. J., Bordas, J., J. Appl. Cryst. *12*, 467 (1979)
50. Boulin, C., Dainton, D., Dorrington, E., Elsner, G., Gabriel, A., Bordas, J., Koch, M. H. J., Nucl. Instr. Meth. *201*, 209 (1982)
51. Gehrke, R., unpublished
52. Vasina, A. A., Gerasimov, V. S., Zheleznaya, L. A., Matyushin, A. M., Sorikin, B., Skrebnitskaya, L. N., Shelestov, V. N., Frank, G. M., Avakyan, Sh. M., Alikhanyan, A. I., Biophysics (USSR) *20*, 813 (1975)
53. Vasina, A. A., Mol. Biol. (USSR) *8*, 242 (1976)
54. Koch, M. H. J., personal communication
55. Chess Newsletter, March 1984
56. Webb, N. G., Samson, S., Stroud, R. M., Gamble, R. C., Baldeschwieler, J. D., J. Appl. Cryst. *10*, 104 (1977)
57. Ameniya, Y., Wakabayashi, K., Hamanaka, T., Wakabayashi, T., Matsushita, T., Hashizume, H., Nucl. Instr. Meth. *208*, 471 (1983)
58. Wakabayashi, K., Hamanaka, T., Anemiya, Y., Tanaka, H., Wakabayashi, T., Hashizume, H., Photon Factory Activity Report 1982/1983
59. Bilderback, D. H., Chess Technical Memorandum No. 16
60. Koch, M. H. J., Bordas, J., Nucl. Instr. and Methods *208*, 461 (1983)
61. Zietz, R., unpublished
62. Bonse, U., Hart, M., in "Small Angle X-Ray Scattering", H. Brumberger Ed., Gordon and Breach, New York (1967)
63. Bordas, M., Koch, M. H. J., Isaac, D., Hart, M., unpublished, cited in Ref. 48
64. Grove, R., Lee, K., Perez-Mendenz, V., Sperinde, J., Nucl. Instr. and Methods *89*, 257 (1970)
65. Gabriel, A., Rev. Sci. Instr. *48*, 1303 (1977)
66. Koch, M. H. J., Bendall, P., Proc. of the Digital Equipment Computer Users Society (1981) 13 DECUS UK
67. Boehm, M., Obermöller, V., Pfeiffer, G., Hoehne, K. H., "Image Management in the System CA-1", Proc. 1st. Int. Conf. of Picture Archiving and Communication Systems, New Port Beach, (1982), SPIE 318 (1982) 161–165
68. Prieske, W., Ph. D. Thesis, University of Hamburg (1984)
69. Bartels, K., Bartunik, H., Prieske, W. Zachmann, H. D., unpublished
70. Forgacs, P., Sheromov, M. A., Tolochko, B. P., Mezentsev, N. A., Pindurin, V. F., J. Polym. Sci., Polym. Phys., Ed., *18*, 2155 (1980)
71. Zietz, R., Heuer, J., Riekel, C., unpublished
72. Grubb, D. T., Liu, J. J. H., Caffrey, M., Bilderback, D. H., J. of Polym. Sc., Polym. Phys. Ed., *22*, 367 (1984)
73. Koch, M. H. J., Bordas, J., Schöla, E., Broecker, H., Ch., Polym. Bull. *1*, 709 (1979)
74. Holland-Moritz, K., Stach, W., unpublished
75. Holland-Moritz, K., Stach, W., Holland-Moritz, I., Progr. Colloid and Polym. Sci. *67*, 161 (1980)

76. Riekel, C., Makrom. Chem., Rapid Comm. *4*, 479 (1983)
77. Riekel, C., Menke, K., Mol. Cryst. Liquid Cryst. *105*, 245 (1984)
78. Elsner, G., Zachmann, H. G., Milch, J. R., Makrom. Chem. *182*, 657 (1981)
79. Elsner, G., Koch, M. H. J., Bordas, J., Zachmann, H. G., Makr. Chem. *182*, 1262 (1981)
80. Zachmann, H. G., Schmidt, G. F., Makr. Chem. *52*, 23 (1962)
81. Prieske, W., Zachmann, H. G. and Koch, M. J., to be published
82. Schultz, J. M., Lin, J. S., Hendricks, R. W., Petermann, J., Gohil, R. M., J. Polymer Sci./Polym. Phys. Ed. *19*, 609 (1981)
83. Fischer, E. W., Report SFB 41, Mainz-Darmstadt
84. Martinez-Salazar, J., Barham, P. J., Chivers, R. A. and Keller, A., to be published, Barham, P. J., Jarvis, D. A. and Keller, A., J. Polymer Sci.-Phys. Ed. *20*, 1773 (1982)
85. Barham, P. J., Keller, A., Otun, L. and Holmes, P. A., J. Materials Sci., in press
86. Spells, S., Barham, P. J. and Keller, A., private communication
87. Fronk, W., Heise, B., Schubach, H. R., Wilke, W., Colloid a. Polym. Sc. *262*, 99 (1984)
88. Gehrke, R., Riekel, C., Zachmann, H. G., to be published
89. Wu, Wen-li, Riekel, C., Zachmann, H. G., Polym. Comm. *25*, 76 (1984)
90. Ornstein, L. S., Zernike, F., Proc. Acad. Sci. Amsterdam *17* (1914) 793
91. Endres, B., Garbella, R. W., Hermann, O., Klemmer, N., Wendorff, J. H., Zietz, R., Koch, M. H. J., HASYLAB-Jahresbericht 201 (1983)
92. Russell, T. P., Hadziioannou, G., Warburton, W., SSRL-Report, Proposal No. 755 M
93. Wegner, G., Bandara, U., Dröscher, M., Frauendorf, B. Zietz, R., HASYLAB-Jahresbericht 207 (1983)
94. Fischer, L., Haschberger, R., Hewel, M., Ruland, W., ibid, 205 (1983)
95. Rothwell, W. S., Martinson, R. H., Forman, R. L., Appl. Phys. Lett. *42*, 422 (1983)
96. Stach, W., Broecker, H. Ch., HASYLAB-Jahresbericht (1983)
97. Caffrey, M., Bilderback, D. H., Nucl. Instr. Meth. *208*, 495 (1983)
98. Holl, B., Heise, B., Kilian, H. G., HASYLAB-Jahresbericht 208 (1983)
99. Forgacs, P., Tolochko, B. P., Sheromov, M. A., Polym. Bull. *6*, 127 (1981)
100. Cabarcos, E. L., Bösecke, P., Zachmann, H. G., to be published
101. Prieske, W., Riekel, C., Zachmann, H. G., to be published
102. Tidick, P., Zachmann, H. G., to be published
103. Wiswe, G., Gehrke, R., Zachmann, H. G., to be published
104. Heise, B., Riekel, C., Stach, W., to be published
105. Chien, J. C. W., Karasz, F. E., Wnek, G. E., Nature *285*, 390 (1980)
106. Prieske, W., Riekel, C., Koch, M. H. J., Zachmann, H. G., Nucl. Instr. a. Methods, *208*, 435 (1983)
107. Zachmann, H. G., Kolloid Zeitschr., v. Z. f. Polymere
108. Riekel, C., Hässlin, H., Menke, K., Roth, S., Mol. Cryst., Liq. Cryst., in press

H. G. Zachmann (Editor)
Received July 5, 1984

# Position Sensitive X-ray Detectors

Jules Hendrix
EMBL, Notkestraße 85, 2000 Hamburg 52, FRG

*This chapter reviews both single photon counting and integrating systems with their respective principles of operation and their properties.*

1 Introduction . . . . . . . . . . . . . . . . . . . . . . . . 60

2 Gas Filled Detectors: Basic Principles . . . . . . . . . . . . . . 60
   2.1 Gas Amplification . . . . . . . . . . . . . . . . . . . . 60
   2.2 Counting Rate Capabilities: The Space Charge Effect . . . . . . . . 64
   2.3 Pulse Shape of Proportional Counters . . . . . . . . . . . . . 66
   2.4 Signal Extraction . . . . . . . . . . . . . . . . . . . . 67

3 Different Types of Position Sensitive Detectors . . . . . . . . . . . 68
   3.1 The Delay-line or LC-line Detector . . . . . . . . . . . . . . 69
   3.2 Detectors with Charge-Division Read-out . . . . . . . . . . . . 71
   3.3 The Backgammon Detector . . . . . . . . . . . . . . . . 73
   3.4 RC-Read-out Detector: The Rise-Time Method . . . . . . . . . 74

4 Area-Detectors: Multiwire Proportional Chambers . . . . . . . . . 75
   4.1 General Principle . . . . . . . . . . . . . . . . . . . . 75
   4.2 Read-out Techniques . . . . . . . . . . . . . . . . . . . 77

5 New Developments: Parallel Electrode Devices . . . . . . . . . . . 80
   5.1 The Principle . . . . . . . . . . . . . . . . . . . . . . 80
   5.2 The Signal Shape . . . . . . . . . . . . . . . . . . . . 81
   5.3 Results . . . . . . . . . . . . . . . . . . . . . . . . 82

6 Photo-Electronic Imaging Devices . . . . . . . . . . . . . . . 83
   6.1 Vidicon-Tube Based Detectors: . . . . . . . . . . . . . . . 83
   6.2 Charge Coupled Devices (CCD), and Photo Diode Arrays (PDA) . . . 89

7 Data Acquisition Systems . . . . . . . . . . . . . . . . . . 91

8 Conclusion . . . . . . . . . . . . . . . . . . . . . . . . 94

9 Appendix: The Derivation of the Expression for the Gas Amplification . . . 95

10 References . . . . . . . . . . . . . . . . . . . . . . . . 97

# 1 Introduction

Synchrotron radiation, as a high intensity X-ray source, has introduced a number of innovations into the field of structural research. X-ray diffraction patterns can be measured with good statistical accuracy within short time intervals.

An efficient collection of the information contained in these patterns is, however, only possible with electronic Position Sensitive Detectors (PSD's).

Time resolved measurements, perhaps the most important application of X-ray synchrotron radiation, are only meaningful with electronic area detectors.

This combination makes it feasible to use X-ray diffraction as a probe to follow structural changes in molecular systems during chemical reactions or phase transitions, in real time.

Position sensitive detectors as imaging devices for ionizing radiation exist for more than a decade in high energy physics and in astronomy. However, the high flux of photons in a two-dimensional pattern demand for new developments, concerning fast positional encoding electronics and highly efficient data acquisition systems. The counting rates in synchrotron radiation experiments differ strongly with the sample under study and can vary from some $10^3$ counts per second up to as high as $10^7$ counts per second.

The large amount of data which has to be collected in time resolved measurements, especially if full two-dimensional patterns are recorded, leads to the development of dedicated high speed data acquisition systems. This is the third component which is necessary together with the source and detectors, to bring synchrotron radiation X-ray experiments to their full deployment.

Among the various detector systems, we distinguish between single photon counting and integrating systems.

In the first case — to which belong the gas filled detector systems — the position of each absorbed photon has to be determined and the result has to be stored in the data acquisition system. In high counting rate experiments, these operations have to be carried out in very short times, in order to minimize dead time losses.

Integrating systems, the main representative of which is the TV-tube detector, do not suffer from this limitation. In these systems, the signals of many photon events are accumulated in the detector itself, and only the integrated result is stored. At first glance, the integrating detectors seem to be advantageous for synchrotron radiation applications. However, several drawbacks have to be taken into account as we will see.

We will describe both types of detectors, with their respective principles of operation and their properties.

# 2 Gas Filled Detectors: Basic Principles

## 2.1 Gas Amplification

The basic principle of operation of gas filled detectors, be it multiwire proportional chambers or linear devices, with a single anode wire, is the mechanism of gas amplification. This kind of detectors have — as its name implies, a filling of an appropriate gas. For the energy range of 5 keV up to 10 keV, a heavy atom gas filling is required, like the noble gases Argon or Xenon with some polyatomic gases added. The role of these admixtures will be discussed later. These noble gases have a high absorption coefficient in this energy range, determining the counting efficiency of the devices [1]. An X-ray in the keV range, produces an initial ion-electron pair by the photoelectric effect. The liberated electron has a kinetic energy equal to the difference between the energy of the absorbed photon (e.g. 8 keV) and the ionization energy of the noble gas (e.g. 25 eV for Argon). This electron will ionize neighbouring atoms by collision, till its initial energy is totally used in ionizations. This means that, in the average, about 300 ion-electron pairs are formed by the absorption of a single 8 keV X-ray. The signal of a single event, with a detector electrode capacity of only 10 picofarad would then have a magnitude

$$V_{Signal} = \frac{n\,e}{C} = \frac{300 \cdot 1{,}6 \cdot 10^{-19}\,\text{Coulomb}}{10^{-11}\,\text{Farad}} \cong 5\mu\,\text{Volt} \qquad (1)$$

where n is the number of primary charges, e the electron charge in Coulombs, and C the capacity of the sensing electrode in Farad. This signal is far too low to be detected directly. It is even very low as compared to the equivalent input noise of modern preamplifiers. For this reason some means of amplification, if possible noiseless is absolutely necessary. In the case of two-dimensional position sensitive detectors, an array of separate amplification elements — one for each picture element — would be needed. This is exactly what is provided in gas filled detectors by the principle of gas amplification. Gas multiplication in a proportional counter is based on the secondary ionization by collisions of electrons with neutral gas molecules. A free electron generated in a gas volume, between electrodes where an electric field is applied, will drift according to the direction of the field lines. If the field is high enough such that, the electron acquires a kinetic energy between collisions, higher than the ionization energy of the neutral gas atoms, secondary ionization will occur, according to Townsend's law.

If $n_0$ is the number of the primary electrons, the increase in the number of electrons, over a path dx in the electric field, will be given by

$$dn = n_0 \alpha\, dx \qquad (2)$$

where $\alpha$ is the first Townsend coefficient, the number of ion pairs formed per cm at normal gas pressure.

By integration, assuming a homogeneous electric field, we obtain:

$$n(x) = n_0 \exp[\alpha x] \qquad (3)$$

The multiplication factor is given by

$$A = \frac{n}{n_0} = \exp[\alpha x] \qquad (4)$$

The Townsend coefficient is a function of the local electric field, and the expression takes the more general form

$$A = \exp\left[\int_{x_1}^{x_2} \alpha(x)\, dx\right] \tag{5}$$

In a proportional chamber, which has a coaxial cylindrical configuration as shown in Fig. 1a, the field has a radial distribution, and is given by,

$$E(r) = \frac{V_b}{\ln\left(\frac{b}{a}\right)} \cdot \frac{1}{r} \quad \text{or} \quad E(r) = \frac{V(r)}{\ln\left(\frac{r}{a}\right)} \cdot \frac{1}{r} \tag{6}$$

with the difinitions as shown in Fig. 1a.

In order to derive an expression for the gas amplification in the proportional chamber type of detector, the integration of the general form (5) has to be carried out for the special case of this radial field distribution.

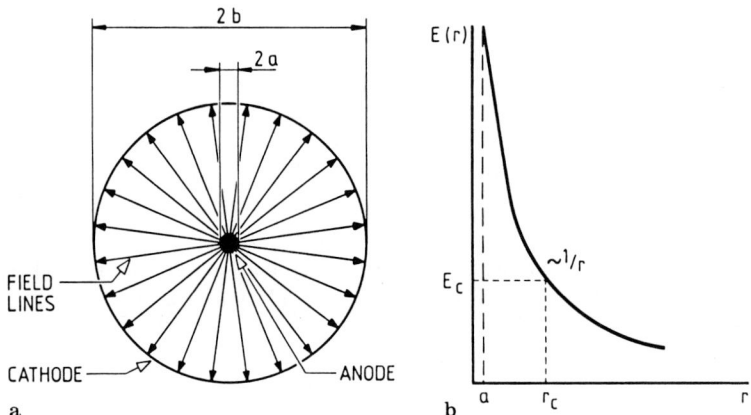

**Fig. 1a.** Proportional counter system. **b** The electric field in a proportional counter

In the appendix, the derivation by W. Diethorn [2] is given, which has been applied successfully to calculate the gas gain of a variety of detector systems with different gas fillings.

Diethorn's expression for the gas amplification A is given by (see Appendix).

$$\ln A = \frac{\ln 2}{\Delta V. \ln\left(\frac{b}{a}\right)} \cdot V_b \cdot \ln\left[\frac{V_b}{Kpa \ln\left(\frac{b}{a}\right)}\right] \tag{7}$$

For a coaxial cylinder, the capacitance is given by

$$C = \frac{2\pi\varepsilon_0}{\ln\left(\frac{b}{a}\right)} \quad (8)$$

where $\varepsilon_0 = 8.8$ pF per meter, the dielectric constant.

Substituting in (7) yields

$$A = \exp\left[\frac{CV_b}{2\pi\varepsilon_0} \cdot \frac{\ln 2}{\Delta V} \cdot \ln\left\{\frac{CV_b}{Kpa2\pi\varepsilon_0}\right\}\right] \quad (9)$$

Now $CV_b = Q$, is the number of changes on the wire per unit length.

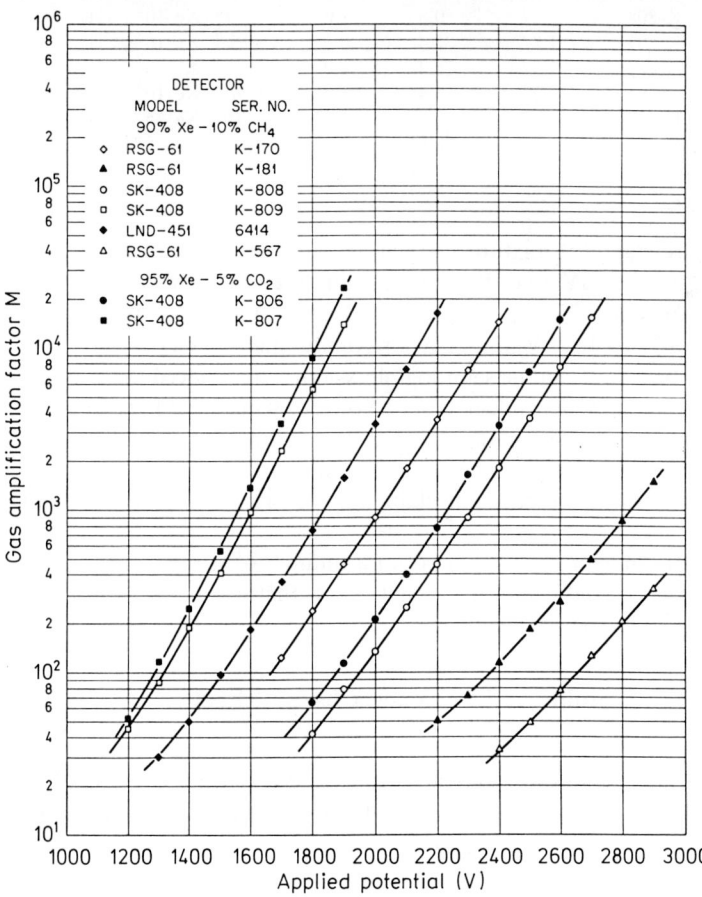

**Fig. 2.** Gas amplification factors for detectors described in Table 1 (From Ref. [3])

Thus, to a first approximation, the gas amplification depends exponentially upon the number of charge per unit length of anode wire, the logarithmic term being of less influence.

The Diethorn expression has been applied in a detailed analysis of commercially available proportional chambers by R. W. Hendricks [3]. Fig. 2 shows the test results for a series of counters, the parameters of which are given in Table 1. The tested devices are not position sensitive detectors, but normal cylindrical proportional counters.

Table 1. Detector Parameters

| Detector Model | Serial No. | b (cm) | a (µm) | p (atm) | Fill Gas |
|---|---|---|---|---|---|
| RSG-61 | K-170 | 2.54 | 25.4 | 1.0 | 90% Xe–10% $CH_4$ |
| RSG-61 | K-181 | 2.54 | 25.4 | 2.0 | 90% Xe–10% $CH_4$ |
| RSG-61 | K-567 | 2.54 | 76.2 | 1.0 | 90% Xe–10% $CH_4$ |
| SK-408 | K-808 | 1.27 | 50.8 | 1.0 | 90% Xe–10% $CH_4$ |
| SK-408 | K-809 | 1.27 | 25.4 | 0.5 | 90% Xe–10% $CH_4$ |
| 452 | 6414 | 1.27 | 25.4 | 1.0 | 90% Xe–10% $CH_4$ |
| 452 | 6415 | 1.27 | 25.4 | 1.0 | 90% Xe–10% $CH_4$ |
| SK-408 | K-806 | 1.27 | 50.8 | 1.01 | 95% Xe– 5% $CO_2$ |
| SK-408 | K-807 | 1.27 | 50.8 | 0.52 | 95% Xe– 5% $CO_2$ |

The gas fillings are standard mixtures, which are also used in position sensitive detectors. The values for $\Delta V$ and K, for the mixtures are $\Delta V = 33{,}9$ V and $K = 3{,}62 \cdot 10^{-4}$ V/atm · cm for the mixture of Xenon and 10% methane and $\Delta V = 31.4$ and $K = 3{,}66 \cdot 10^{-4}$ V/atm · cm for Xenon and 5% $CO_2$.

The almost linear behaviour on the semi-log plots shows that indeed the amplification is roughly exponentially dependent on the voltage, with the other parameters unchanged. A comparison between the graphs for the Ser.-K-808 and Ser.-6414 counters shows the effect of changing the anode wire diameter. The curves for K-806 and for K-807: show the effect of changing the pressure, leaving the other parameters unchanged.

Although the Diethorn expression is based on a number of strong simplifications in the model for the gas amplification, the results show that it is perfectly usable for the prediction of the properties of gas filled detectors.

## 2.2 Counting Rate Capabilities: The Space Charge Effect

The derivation for the gas amplification does not take into account the effect of the drifting positive ions in a proportional counter. Indeed, due to the narrow amplification region, around the anode wire, practically all the charges are created in a very small volume. These ions drift rather slowly away from the wire. They tend to reduce the electric field around the wire. If a second photon is detected at the same position on the wire, the field and consequently, the gas amplification will be lower. Since the density of the positive ions, in the anode-cathode gap increases with the local count rate, the pulses tend to become smaller. Usually a minimum signal amplitude is required for processing. At higher pulse rates, counts will be lost, due to the decrease in gas amplification.

The general problem of the space charge effect in pröportional tubes has been analysed by R. W. Hendricks [4] who derived an analytical formulation.

The sheath of positive ions causes the effective potential of the anode wire to decrease.

The attenuated gas amplification of the counter is given by:

$$A = \exp[K(V_b - \delta V)] \tag{10}$$

where $\delta V$ is to change in the potential due to the space charges.

As compared to the amplification at very low count rates $A = \exp[KV]$, the gain is reduced by a factor:

$$\frac{A}{A_0} = \exp[-\delta V] \tag{11}$$

The decrease in the potential is given by Hendricks as

$$\delta V = \frac{n_0 A_0 p \ln\left(\frac{b}{a}\right) b^2 R}{2\pi \mu_+ V_b L} \tag{12}$$

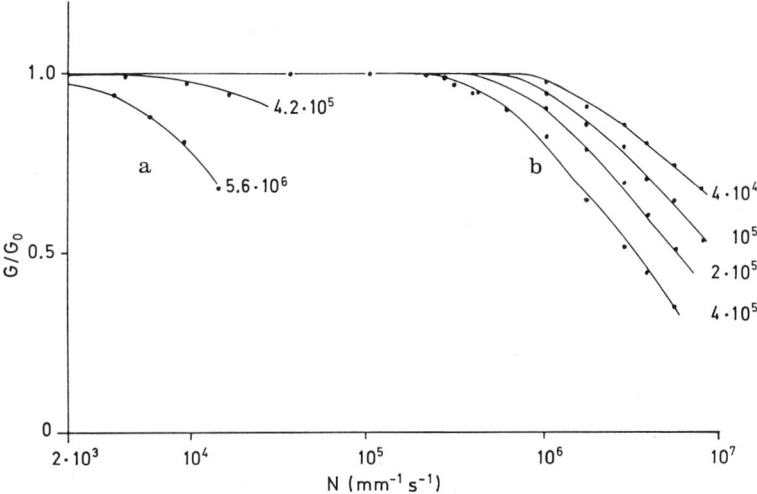

**Fig. 3a and b.** Pulse height (normalized to low rate pulse height) versus count rate per anode wire length. Parameter: the number of charges per avalanche. **a** Standard proportional counter; **b** Optimized high rate counter (From Ref. [5])

where $n_0$ is the number of primary electrons, $A_0$ is the gas gain, $n_0 A_0$ = number of charges per avalanche, $\mu_+$ is the drift mobility of the positive ions, $V_b$, the counter voltage, L is the counter length, R is the count rate, per second and per unit length of anode wire and p is the gas pressure in Torr. Obviously, b is the most important parameter to be adjusted. This means, for high counting rates, the anode-cathode gap should be made as small as feasible. The change in potential is proportional to $A_0$, the gas amplification at low count rate. For high event rates, it is advantageous to work at lower gas gains and higher external electronic amplification. This implies, however, the use of low noise amplifiers. The dependence of $\delta V$ on L, and the fact that positive charges drift away perpendicularly from the anode wire, explain why a proportional counter can be "dead" locally, if local intensities become high. In real measurements, this can lead to completely wrong results, if a strong reflection increases. Due to the diminuation of the gas amplification, counts are lost and the peak appears to decrease.

Optimization of all parameters involved, can result in a significant increase of counting rate capability. As an example, results obtained with a test chamber, and carefully designed low noise electronics are shown in Fig. 3, from Ref.[5]. An increase of the counting rate of two orders of magnitude has been obtained with an optimized high rate chamber.

## 2.3 Pulse Shape of Proportional Counters

The time dependence of the signal in proportional chambers has been derived by several authors[6,7,8]. The signals detected with proportional chambers, are a consequence of the change in energy of the system, due to the movement of charges. A detector is in fact a capacitor, where charges are generated and moved between its electrodes. The movement of primary charges does not contribute measurably to the signal. The electrons resulting from the multiplication near the anode wire are collected after travelling through a small potential difference, so they contribute only minimally to the signal.

Most of the signal is due to the sheath of positive ions, moving over the whole potential difference across the chamber:

The current pulse shape is given by:

$$i(t) = \frac{n_0 A}{2 t_0 \ln\left(\frac{b}{a}\right)} \cdot \left(1 + \frac{t}{t_0}\right)^{-1} \tag{13}$$

and the charge:

$$Q(t) = \frac{n_0 A}{2 \ln\left(\frac{b}{a}\right)} \cdot \ln\left(1 + \frac{t}{t_0}\right) \tag{14}$$

where:

$$t_0 = \frac{a}{V_b} \cdot \ln\left(\frac{b}{a}\right) \cdot \frac{a}{2\mu_+} = \frac{a}{2\mu_+ \cdot E_a}$$

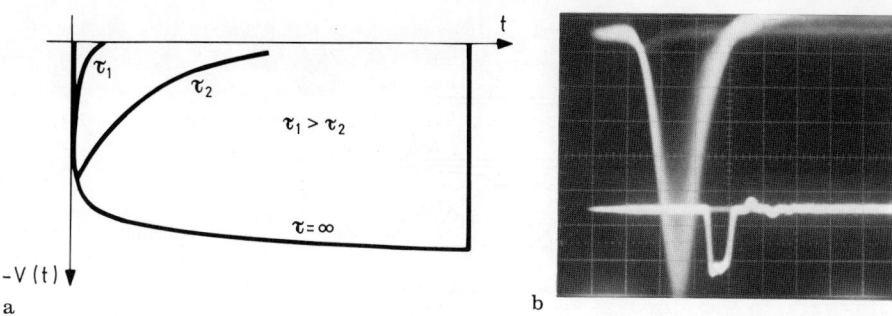

**Fig. 4a and b.** The signal shape of proportional counters. **a** Schematically; **b** Typical pulse shape. (hor.: 20 ns/div; vert.: 20 mV/div)

$\mu_+$ the ion mobility is 1,7 cm²/sec V atm in a typical counter mixture containing 90% Argon and 10% $CO_2$.

Since $E_a$ is very high in the vicinity of the anode wire (some $10^5$ V/cm), the signal rises very rapidly, and then grows slowly until the positive ions reach the cathode. This can take up to several hundred μsec.

The pulse shape depends strongly on the detector parameters like the anode diameter and the applied voltage.

Normally, the signal is differentiated in the preamplifier circuit. Typical signal shapes are shown in Fig. 4 for different differentiating time constants T = RC.

## 2.4 Signal Extraction

In the previous section, it was shown that the signals, generated in a MWPC-system are due to the movement of the positive ions of an avalanche in the electric field. The information to be retrieved is the position in one or two dimensions of the impact of the photon.

If the signal could be measured without noise, it would be possible with any method to define the position of an avalanche. But, gas amplification is a statistical process, and the signals underly fluctuations in amplitude, which have to be taken into account in the read-out methods.

Although, gas multiplication factors as high as $10^3$ to $10^5$ can be achieved, the signal charges are of the order of $10^6$ electrons or only $10^{-13}$ coulomb. Consequently, some electronic preamplification has to be installed before the signals can be processed.

Preamplifiers unavoidably add electronic noise to the signals of the detector. It is of a crucial importance for the development of PSD-systems to make the right choice concerning the preamplifiers. In fact, the art of designing a good detector system lies in the art of designing the best preamplifier for a given purpose.

A number of publications give a good overview of the basics of the optimization of signal processing systems for timing and amplitude measurements [9,10,11].

In the following section we will discuss several read-out methods of position sensitive detectors. They rely either on the accurate measurements of charges or of time differences between signals.

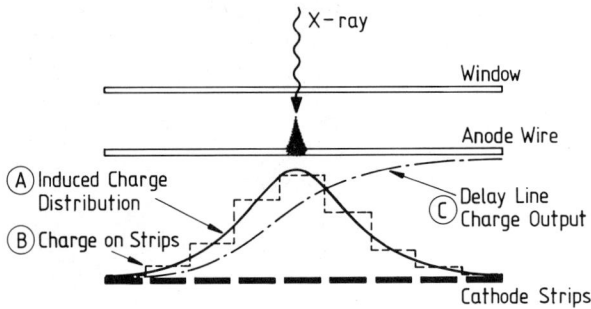

Fig. 5. Schematic of the induced signal in a position sensitive detector. (From Ref. [13])

In general, the spatial resolution of PSD's is a function of the signal to noise ratio after the preamplifiers. Since the noise amplitude is proportional to the bandwidth of the amplifier stages, it is absolutely necessary to select this bandwidth such that the maximum information, timing or amplitude, is extracted from the signal, and that the minimum amount of noise is passed.

A fundamental contribution concerning the optimization of filtering for position sensitive read-out system: is made by V. Radeka of Brookhaven National Laboratory [12].

It was already pointed out in the previous sections that a PSD is in fact a capacitor system. The movement of charges between the electrodes of the system creates a displacement current in the system. The induced signals have a lateral distribution, depending on the geometrical configuration of the PSD.

In multi-wire proportional chamber systems (MWPC-systems), and also in linear position sensitive detectors, the cathode electrodes are usually divided up into individual strips, from which the positional information can be extracted.

Fig. 5 from Ref. [13] depicts schematically how the induced charges are distributed in such a system.

As indicated in Fig. 5, the signal is spread out over several elements of the read-out electrode. The role of a position sensitive read-out system is to determine the centroid of the signal distribution.

Curve B shows the portion of the signal collected by each strip. In a delay-line read-out system, the individual strips are connected to consequent taps. The signal, made up of the individually induced charges, propagate through the delay-line and the resulting signal output is given schematically by curve C.

As shown by Gatti [14], it is mandatory to exactly determine the optimal dimensions of the read-out strips, if a high spatial resolution has to be obtained. However, in many practical systems the design is prescribed by the dimensions of components such as connectors for instance. A substantial loss in spatial resolution can be the consequence.

## 3 Different Types of Position Sensitive Detectors

The basic principles of operation which have been explained so far, are valid for one dimensional or linear detectors as well as for two-dimensional gas filled detector systems.

# Position Sensitive X-ray Detectors

In all cases, the aim is to determine the position of the centroid of the induced signal with the highest resolution possible. The various types of position sensitive detectors differ only in the way in which the localisation of a photon absorption event is achieved.

Thus, a discussion of the different systems is in fact a discussion of the read-out techniques which are applied to obtain the positional information.

## 3.1 The Delay-line or LC-line Detector

The principle of operation of this kind of detector is shown in Fig. 6. The delay-line read-out technique is simple, at least in principle, and needs a minimum of electronic components, most of which are commercially available, in the form of standardized NIM-modules.

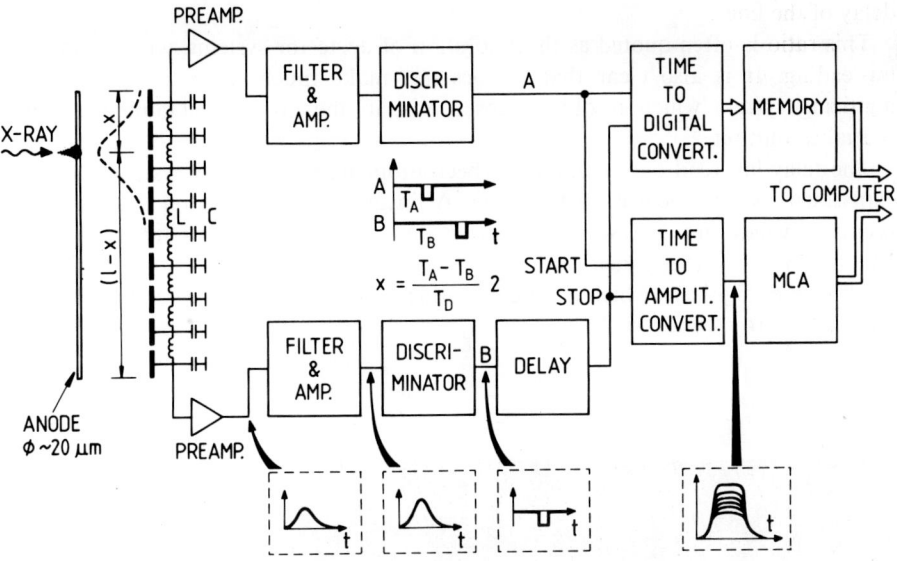

**Fig. 6.** Principle of the delay-line detector

The technique was first introduced in 1968, as a read-out system for a two-dimensional MWPC for high energy physics experiments [15]. Basically, the signal induced in the read-out strips propagates towards both sides of the delay-line, and the difference in time of arrival at both ends is a measure for the position of the avalanche.

Radeka derived an expression for the relative spatial resolution which can be obtained with this technique (12):

$$\frac{\delta l}{l} = 1{,}2 \cdot \frac{v_n}{Q_S} \cdot \frac{1}{Z_{DL}} \cdot \frac{t_r^{3/2}}{T_{DL}} \tag{15}$$

where:

$v_n$ = the rms equivalent input noise of the preamplifier, in Volts per Hz;
$Q_s$ = the signal change in coulambs;
$Z_{DL}$ = the impedance of the delay-line in Ohm;
$T_{DL}$ = the total delay over the length of the line;
$t_r$ = the rise time of the signals in the delay-line.

The first term expresses that the expected resolution is inversely proportional to the signal to noise ratio of the system. The second term means that the result improves if better use is made of the signal by using a higher impedance of the line. And the third term explains how the resolution is dependent on the timing properties of the delay-line.

From the expression it can be concluded that the spatial resolution is not merely equal to the ratio of the time resolution of the time digitizer to the total propagation delay of the line.

This ratio is often quoted as the resolution of dectector systems, which is at least misleading. It is also clear that it does not make much sense to select a time digitizing system which resolves picoseconds, in the hope of obtaining a better spatial resolution.

The delay-line read-out technique has been adopted to a number of systems for X-ray experiments with synchrotron radiation. A linear detector based on this technique is currently used in time resolved measurements on polymer systems at the X13 beam line at EMBL [16]. (See other Section.)

The time digitizer, which is used in this set-up, is a commercially available module (Lecroy, Model 4122).

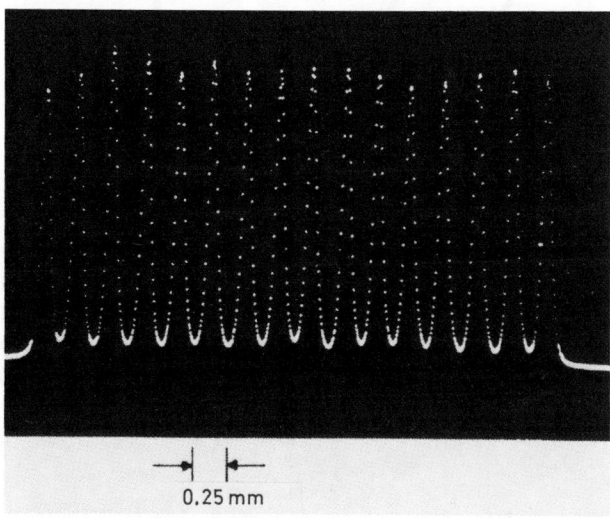

**Fig. 7.** Example of spatial resolution with delay-line read-out. (From Ref. [17])

The delay-line read-out is probably the most popular technique. The necessary electronic functions are all commercially available, including some time digitizers.

A spatial resolution as good as 0.1% can be achieved, if all components are carefully selected.

Figure 7 shows the result of a spatial resolution test with a delay-line read-out [17]. The result was obtained with a Xenon-$CO_2$ (90%–10%) gas-filling, at normal pressure. The width of the peaks in 70 µm FWHM. The delay-line in the test had an overall delay of 1 microsecond and an impedance $Z_{DL} = 500\ \Omega$. The signal charge was about 0,6 picocoulomb.

The counting-rate capability of this type of detector is determined by the total delay of the line, which is usually between 200 nsec and 1000 nsec, and by the conversion time of the processing electronics, which is of the order of 1 microsecond in the case of a direct time digitizer.

The peak count-rate is thus $5 \cdot 10^5$ events $\sec^{-1}$, leading to about $5 \cdot 10^4$ counts $\sec^{-1}$, with negligible dead-time losses.

The classical time measuring system, the combination of a Time-to-Amplitude Converter (TAC) and a Multi-Channel-Analyser (MCA) has a much longer dead-time of the order of 20 µsec.

The only disadvantage in synchrotron radiation applications is the apparent counting rate limitation due to the propagation delay in the line. But, a comparison of this propagation delay with charge collection times in other methods is still advantageous for the delay-line method.

## 3.2 Detectors with Charge-Division Read-out

This method is shown schematically in Fig. 8. From the principle, it seems to be the simplest method of all. But, to obtain a good positional resolution, a very careful design of the processing electronics is necessary.

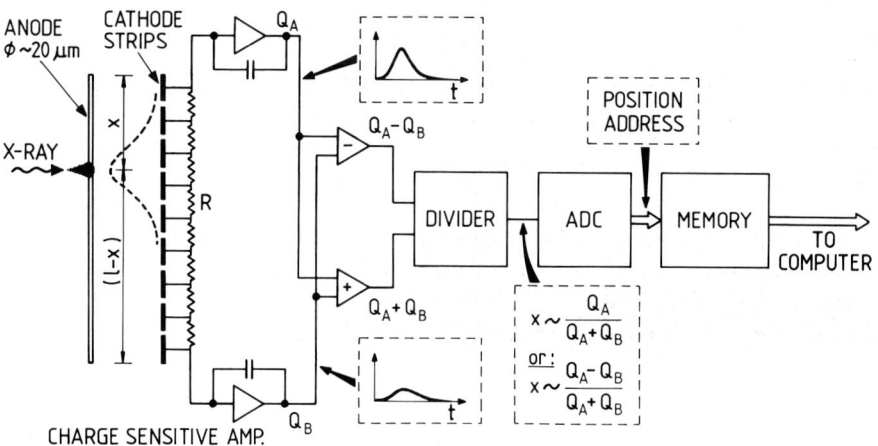

**Fig. 8.** Charge division read-out

The resistive chain, shown in Fig. 8, can be a resistive anode wire itself, for instance in a linear position sensitive detector, or it can consist of resistors connecting the cathode read-out pads.

The method is based on the division of the signal charge over both sides of the resistive chain. The charges, as seen by the preamplifiers, are inversely proportional to the impedances from the position of the avalanche to each end of the detector. With $Q_S$, the total charge of the avalanche, and the definitions of Fig. 8, we obtain:

$$Q_A = \frac{Q_S}{x \cdot \varrho} \quad \text{and} \quad Q_B = \frac{Q_S}{(l-x) \cdot \varrho} \tag{16}$$

where $\varrho$ is the resistivity per unit length of the resistive chain.

The position, x, can be determined from the relationship:

$$\frac{Q_A}{Q_A + Q_B} = \frac{x}{l} \tag{17}$$

The best resolution which can be achieved with optimal filtering does not depend on the value of the resistive chain R, but solely on the eletrode capacitance. The optimum resolution is given by [12]:

$$\frac{\delta l}{l} = 2,5 \cdot \frac{(kTC_D)^{1/2}}{Q_S} \tag{18}$$

$k$ = Boltzmann's constant
$T$ = absolute temperature (°K)
$C_D$ = the electrode capacitance (see Fig. 9)
$Q_S$ = signal charge (Coulomb)

In order to achieve a resolution $\delta l/l = 0.1\%$, a signal about ten times higher than with the delay-line read-out is necessary.

Since the resistive chain together with the capacitances C behaves also as a diffuse RC-line, it takes some time for the charges to reach the ends of the detector. The resolving time — this is the time over which the charges have to be collected — has to be not much less than $T_D = R_D C_D$, the detector time constant, where:

$$R_D = R \cdot l \quad \text{and} \quad C_D = C \cdot l.$$

with R and C the resistance and capacitance per unit length. Since $T_D = RCl^2$, the resolving time increases with the square of the length. For practical systems T can be as short as 50 nseconds.

The counting rate is not solely limited by the charge collection time. In addition to that, the devision $Q_A/(Q_A + Q_B)$ has to be executed either by an analog divider or by a digital processor. Both methods are not trivial tasks, since a high accuracy has to be maintained over the full amplitude range of the signals. In both cases, the analog charge signals have to be converted into a digital value before or after the division.

These operations are quite time consuming, even with modern components resulting in counting rate limitations which are below those of the delay-line method at equal spatial resolution.

## 3.3 The Backgammon Detector

An alternative to the resistive chain for charge division of the avalanche signal can be achieved in an elegant way by a geometrical arrangement of the read-out electrode [18]. The geometry of the cathode has the form of interleaved triangles, like the well-known backgammon game as is shown in Fig. 9a. The charge, as seen by each amplifier is proportional to the area of the cathode elements exposed to the induced signal. In the backgammon arrangment, this area is linearly dependent on the position of the avalanche.

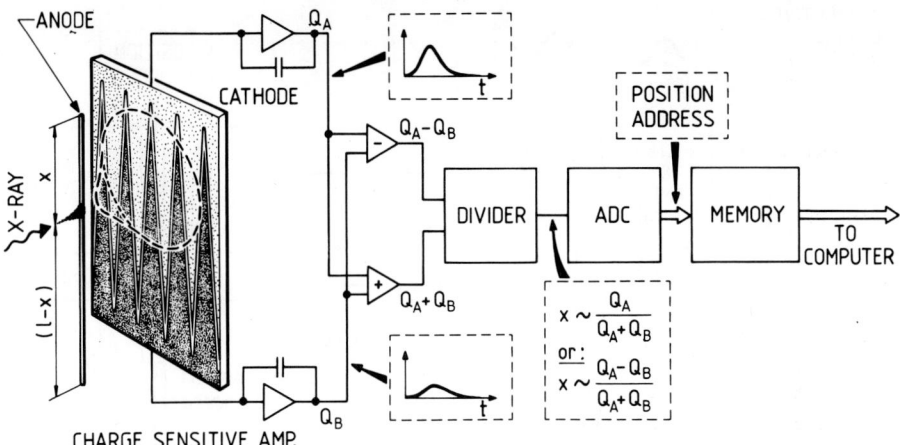

**Fig. 9.** The backgammon detector

The spatial solution is the same as with the resistive charge division method and is given by expression [24]. The resolution is solely a function of the capacitance of the read-out elements, and so there is no fundamental improvement to be execped as compared to the resistive charge division. However, the elegant read-out method has the advantage that the read-out plane can be produced with high mechanical accuracy, and that a minimum number of components are necessary.

The same count rate limitations as with the previous method apply due to the charge collection time, the determination of the ratio of the signals and the analog-to-digital conversion.

A device, based on this principle, is commercially available.

The spatial resolution is 186 µm FWHM, for a 60 mm detector, corresponding to a $\delta l/l = 320$, with a 1 atmosphere Xenon-$CH_4$ (90%, 10%) filling. This result was obtained at an amplification factor of 6000, such that the signal charge with 8 keV radiation is 0.3 pCoulomb [18].

The maximum counting rate is specified as 15 000 counts $sec^{-1}$, due to the analog division and the subsequent analog-to-digital conversion.

## 3.4 RC-Read-out Detector: The Rise-Time Method

As compared to the previous method, this technique uses the resistive chain as a component of a diffuse transmission line [22]. The resistive elements are replaced by a highly resistive anode wire which, together with its capacitance per unit length, forms a distributed RC-transmission line. The anode wire can be made as a quartz fiber coated with carbon.

In fact, in this method there is no necessity of any other read-out electrode, as for instance in the case of the delay-line technique. This can be certainly an advantage. But unluckily, the highly resistive anode wire shows to be very sensitive to mechanical damage. An occasional high voltage breakdown either destroys the coating or, at least, changes its properties locally and thus the detector response.

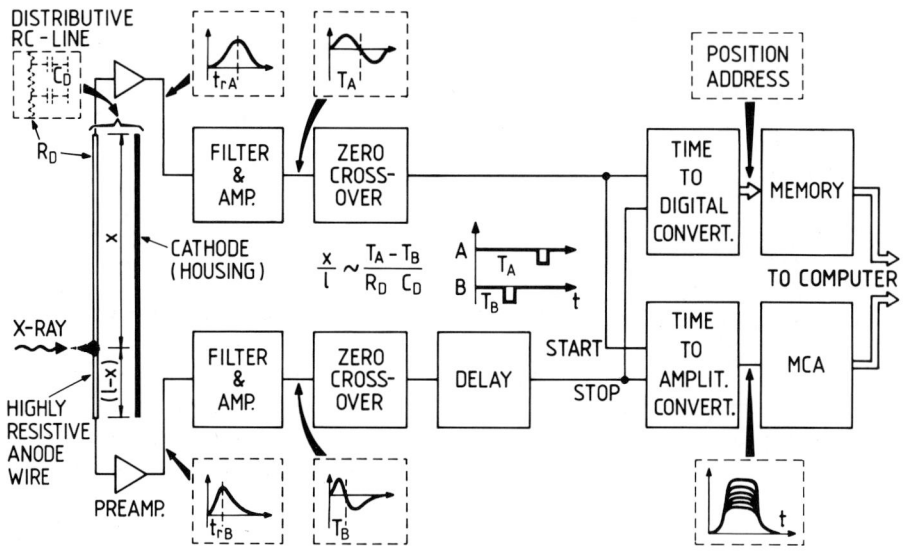

**Fig. 10.** The RC- or Rise-Time read-out

The RC-method, or rise-time method, is shown in Fig. 10. Charges injected into the wire by an avalanche at a location x cause a current to flow through the anode wire in both directions, until the initial charge is spread out over the entire length. The position of the avalanche is estimated by measuring and comparing the two currents at both ends of the detector.

One can show that by appropriate filtering of the signals [22]

$$x = \frac{\Delta t_c}{S} \qquad (19)$$

where $\Delta t_c$ is the time interal between the zero crossover at the left and the right side.

The proportionality contant S is given by
$$S = RC_t$$
with
$$C_t = Cl + 2C_{load}$$
C and R are the distributed capacitance and resistance per unit length of the RC-line.

The positional resolution is inversely proportional to the sensitivity constant $S = RC_t$, which means that the resistivity of the wire has to be high. Again, a choice has to be made between a good spatial resolution and a high counting rate capability.

This method has been applied successfully in the read-out of one-dimensional [19,20] and two-dimensional PSD's [21].

Some commercially available linear detector systems use this rise-time technique. The resolutions that are specified are 50 μm FWHM, at a signal charge of 0.5 pico-Coulomb, for an overall detector length of 50 mm ($\delta l/l = 10^{-3}$).

The countrate capability of these devices is dictated by the signal processing electronics and amounts to about $10^4$ counts sec$^{-1}$ maximum.

# 4 Area-Detectors: Multiwire Proportional Chambers

## 4.1 General Principle

G. Charpak and his collaborators introduced in 1968 the multiwire proportional chambers (MWPC) as 2-dimensional position-sensitive detectors in high-energy physics experiments [22]. These detectors are presently being used more and more in X-ray scattering and -diffraction experiments. A MWPC consists of a plane of fine, equally spaced parallel anode wires, positioned between two cathode planes. The latter can be wires themselves, or conductive planes. The cathode planes usually are configured as read-out elements for the localisation of the X-ray absorption events. Figure 11 shows schematically the general set-up of a MWPC. The electric field configuration around the anode wires is shown in Fig. 12 from Reference [23]. We recognize the radial appearance of the field lines in the vicinity of the anode wires. It becomes clear that, in fact, each anode wire, with the neighbouring cathode elements, acts as an individual proportional counter. Indeed, most of the derivations shown above for the coaxial proportional chamber are valid for each of the linear cells in a MWPC. Erskine [23] has derived expressions for the field in such a system, and also for the capacitance per unit length of the anode wire with respect to the outer cathodes.

Table 2 shows the values as calculated for two anode-cathode gap thicknesses, anode wire diameters (2a) and spacings between them (s) (see Fig. 11).

From the derivation for the gas amplification in section II-A, we know that to a first approximation, A is a function of $CV = Q$, the number of charges per anode length. From the values in Table 2 it becomes clear that, if the pitch s is made smaller, the capacitance decreases. This means that the operating voltage V, has to be increased to maintain a sufficient gas amplification. Another possibility is to choose finer anode wires.

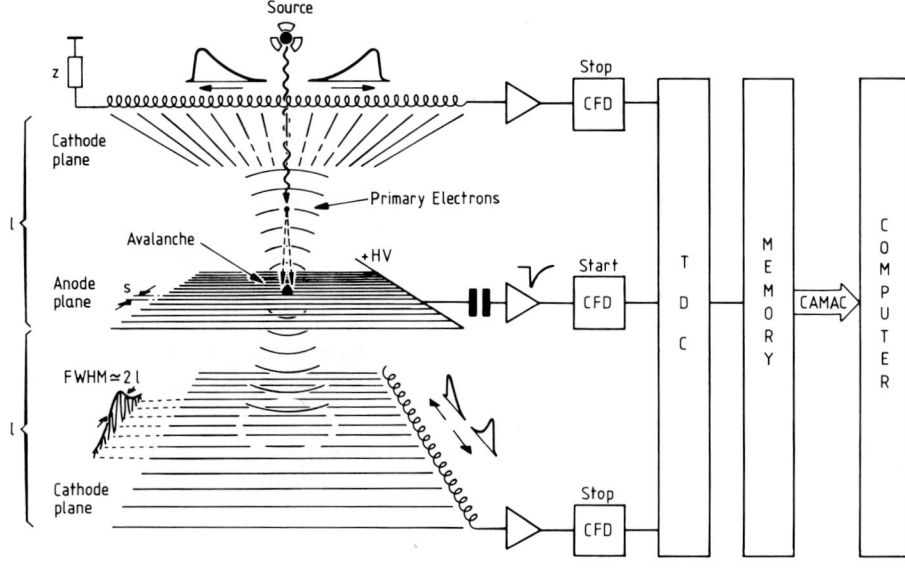

Fig. 11. Schematic of a multi-wire proportional chamber

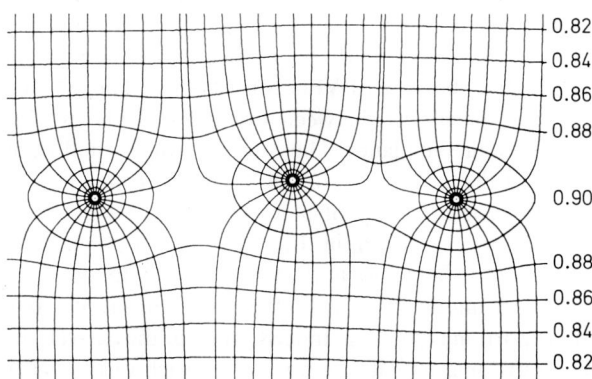

Fig. 12. The electric field in a multi-wire proportional chamber. (From Ref. [23])

In practice, 20 μm wires are rather easy to handle, but 10 μm wires become very fragile and an accidental high voltage breakdown unavoidedly breaks these thin wires. So, practically, one is limited to minimum wire distances of approximately 1 mm.

Since the avalanches are generated in the close proximity of anode wires, the spatial resolution of MWPC will be quantized in the direction perpendicular to the anode elements. This is one of the principal disadvantages of the MWPC as a position sensitive detector.

**Table 2.** Capacitances, in picofarads/metre, of a wire in the central plane with respect to the outer electrodes (From Ref. [23])

| mm | $2r_a$ (μm) | C (pF/m) | | |
|---|---|---|---|---|
| | | s = 3 mm | s = 2 mm | s = 1 mm |
| 7 | 20 | 4.97 | 3.85 | 2.25 |
| | 30 | 5.21 | 3.96 | 2.28 |
| | 40 | 5.30 | 4.04 | 2.31 |
| | 50 | 5.41 | 4.11 | 2.34 |
| 3 | 20 | 7.94 | 6.81 | 4.56 |

The spatial resolution along the anode wires, on the other hand, can be very good in a MWPC. As we will see, there exist several positional read-out schemes which allow a better resolution than the distance between the cathode read-out elements. Perpendicular to the wires, the resolution is limited to the distance between them, since there is no signal generation possible, except on the anode wires.

MWPC-devices have been used successfully in X-ray diffraction work, in recent years. In synchrotron radiation experiments, however, it will not be so much the detector, but more the read-out electronics which will not be able to deal with the enormous event rates, which are to be expected in two-dimensional patterns.

## 4.2 Read-out Techniques

The various read-out techniques, as described before, for linear PSD's can be applied for two-dimensional systems too. One should however bear in mind that, when complete diffraction patterns have to be measured, the conting rate capability plays an even more important role. A linear detector is exposed to only a fraction of the total flux of a pattern, as compared to a two-dimensional detector, which is intended to measure the entire area at once.

Although the local count rate is limited by the space charge effect, as we have seen before, it is the read-out system which will be the limiting component of the system as a whole. Certainly with synchrotron radiation as a source, but already with a rotating anode X-ray generator, area detector systems will be saturated quickly. In addition, the demands concerning the data acquisition system become even more stringent.

In some cases it is advisable to tailor the detector to the need of the application, in order to minimize the number of read-out channels.

For instance, in small angle scattering experiments, only the angular distribution of the intensity in the centro-symmetrical pattern is wanted. A circular detector, with concentric ring-shaped read-out elements, directly measures this distribution. Instead of measuring the entire pattern in two dimensions, the special geometry of the detector reduces the information to ring-shaped resolution elements. In practical terms, this means that a detector with 256 ring-shaped resolution elements, replaces a 256 × 256 element fully two-dimensional system. This significant reduction of the number of read-out elements evidently reduces the number of channels of the data aquisition system. Such a ringe-shaped detector system, divided up into quadrants, is used successfully at the EMBL Outstation in Hamburg for time-resolved small angle scattering measurements of biochemical solutions [24].

In many applications full two-dimensional information is necessary, and thus fast read-out systems have to be provided. The various techniques, which have been discussed before, have been applied with area detectors. The delay-line read-out has been used with two-dimensional detectors in synchrotron radiation laboratories. At the DORIS-ring in Hamburg an area detector with a 1 mm anode wire distance and a total area of 200 mm × 200 mm is currently in use for measurements of muscle diffraction patterns and for X-ray crystallography [25]. The spatial resolution is about $(2,5 \times 2,5)$ mm$^2$ FWHM.

The count rate at 10% dead time loss is as high as $2.10^5$ sec$^{-1}$. A similar detector is used for X-ray crystallography at the University of California [26]. The spatial resolution is reported as 750 µm × 2 mm FWHM. The counting rate capability of this device is $5.10^4$ sec$^{-1}$ at 10% dead time loss.

At SRS at Daresbury also the delay-line read-out method has been chosen [27]. The detector system with an area of 200 mm × 200 mm has a resolution of about $(2 \times 2)$ mm$^2$. The count rate capability is given as $1,5 \cdot 10^5$ sec$^{-1}$.

In all three cases, the main application of area detectors is X-ray crystallography.

The rise-time technique has also been applied to two-dimensional detectors. A spatial resolution of 2 mm by 1 mm FWHM has been reported [21]. As with linear detectors, the count rate capability of this method is lower than with the delay-line read-out, making it less useful for synchrotron radiation experiments.

If a high spatial resolution is needed from large area detectors, some degree of segmentation can be useful. This is not possible with the delay-line method, without discontinuities between sections. In the case of charge division, a very useful scheme has been proposed, which could result in a submillimeter spatial resolution.
The method uses the charge division technique for interpolation over only one section of the total detector, at the time. There is one charge sensitive amplifier per section, increasing the number of electronic components. This is only a minor disadvantage, since the gain in positional resolution is $N^{3/2}$, where N is the number of sections.

A more advanced read-out method, which also depends on segmentation, is the so called centroid finding method [28]. Spatial resolutions of the order of 0,1% have been reported. The method is however quite involved concerning the processing electronics and has been applied only to a neutron detector at present [29]. The centroid finding technique is certainly very promising for the application in synchrotron radiation experiments.

The most elaborate method concerning the number of electronic components but, on the other hand, by far the simplest is the "wire-per-wire" read-out. This method, for the two dimensional case, is depicted in Fig. 13.

In this method, the signal originating at each wire (both the anode wires and the cathode wires) is processed by an individual amplifier-discriminator circuit. This requires many amplifiers. For instance, for a two-dimensional detector with a resolution of 256 × 256, a total of 512 amplifier-comparator systems are necessary. Each circuit on its own, however, is not complicated, and large scale fabrication is not too difficult.

Figure 13 shows the general principle of the method [30]. The determination of which anode wire has been hit by an avalanche is obvious. An avalanche at an anode wire

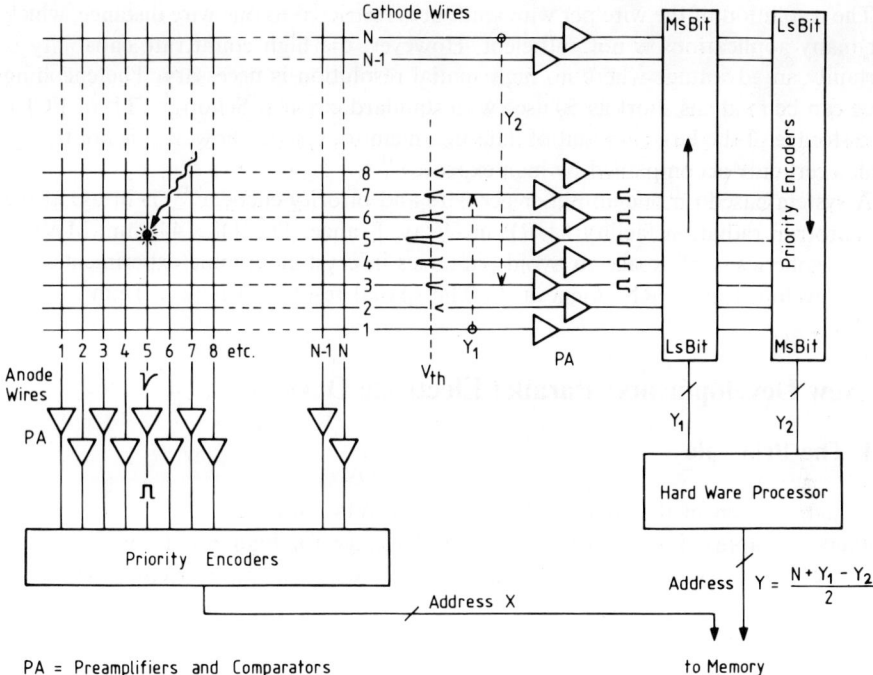

**Fig. 13.** The principle of the wire-per-wire read-out

produces a signal higher than the threshold voltage at the comparator. The latter produces a signal of logical "one" level at the input of a priority encoder chain. This device encodes the address of the anode wire hit. This address can be latched for later processing. The encoding of the anode wire is done very rapidly (typically 20–50 nsec). For this reason, the wire per wire method is well suited for very high counting rates.

The readout of the cathode plane is somewhat more complicated, since the induced signals extend over several elements. The determination of the position of the avalanche is done by the so-called "center of cluster" method. The thresholds of the discriminators are chosen such that the induced signal of several cathode wires switches them (see Fig. 13). Two priority encoder systems are connected to the cathode plane in opposite directions. Priority encoders produce the address of the highest numbered input element which receives a signal. Since the systems are connected in opposite directions, system A will deliver the address of the higher numbered cathode wire whose signal is above the threshold value. System B gives the address of the lower wire.

The wide spreading of the induced signals (see Fig. 5) makes the trimming of the threshold voltages of the comparators very critical. Slight deviations in the thresholds, of the order of 1% result in differential linearity fluctuations of 10%. These again manifest themselves as a modulation of the sensitivity over the detector plane of the same magnitude. This is the weakness of this method, which makes it less suitable for accurate measurements.

The resolution of the wire per wire readout is restricted to one wire distance, which for many applications is not sufficient. However, the high count rate capability is certainly an advantage where no high spatial resolution is necessary. The encoding time can be made as short as 50 nsec with standard circuits (Schottly TTL or ECL). The storage of the large amount of data in a memory systems however, is not trivial, and is certainly accompanied by high expenses.

A system based on one amplifier per wire and priority encoder [31] is in use at the synchrotron radiation facility LURE in Orsay, France. The $448 \times 448$ mm MWPC, with a spherical drift space to avoid paralaxes in crystallographic experiments, has been developed at CERN, Geneva [32]. The spatial resolution is $(2 \times 2)$ mm².

## 5 New Developments: Parallel Electrode Devices

### 5.1 The Principle

The quantization of the spatial resolution of MWPC-Systems, with minimum wire distances of approximately 1 mm, is a disadvantage for high resolution work, for instance, X-ray crystallography. Recently new developments have been undertaken,

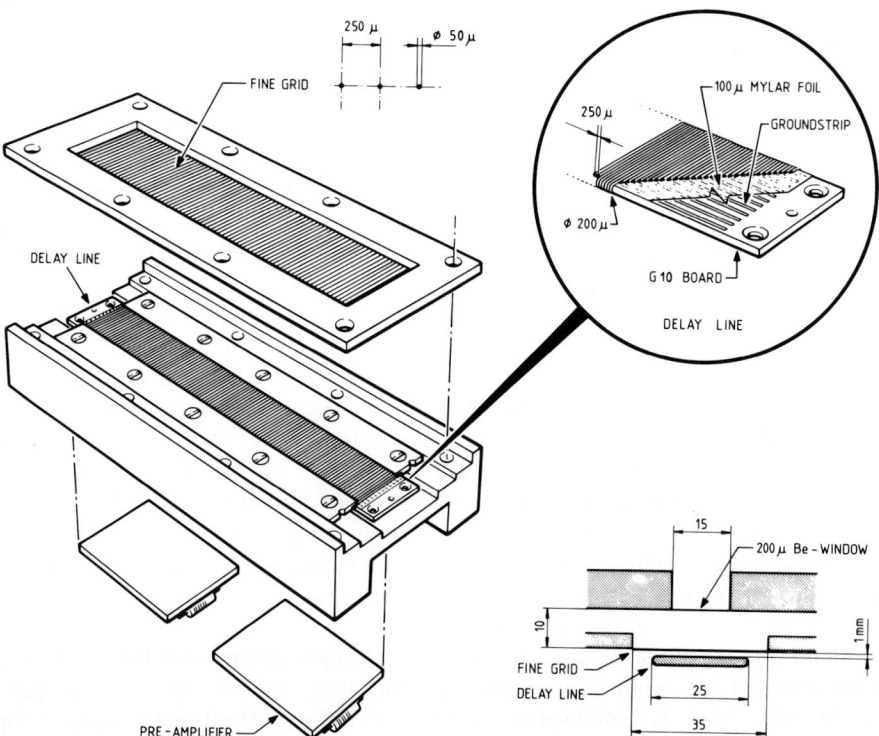

Fig. 14. Schematic of a parallel electrode position sensitive detector: The fine grid detector

with the aim to overcome this limitation. The new systems are parallel electrode devices, based on gas amplification in uniform electric fields.

The general structure of such a devices is shown in Fig. 14 from Ref. [33].

The upper electrode, called fine grid in Fig. 14, consists of 50 µm copper-berylium wires at a distance of 250 µm. The lower electrode — in this case the delay line — is the positively charged electrode or anode. The detector is devided into two gaps by the fine grid. The upper 10 mm wide absorption gap, where the electric field is low, and the narrow (1.5 mm) amplification gap, where the field is about 20 kV/cm. The aim of the fine structure of the grid, is to minimize the modulation of the electric field in the amplification gap. Primary electrons, formed in the upper gap by the absorption of X-rays, are swept into the amplification gap. Here the field over the entire gap depth, is sufficiently high to cause secondary ionization. The important difference with a MWPC-structure is that, the field is not concentrated around single wires, but is constant over the entire amplification region. Due to the fine structure of the fine grid, the individual wires do not play a role in the gas amplification. The grid behaves as if it were a continuous electrode, which is transparent for electrons.

As a consequence, signal generation is possible, continuously over the entire area of the detector, independently of the microstructure of the electrodes.

## 5.2 The Signal Shape

The growth of an avalanche in a sufficiently high uniform electric field is given by Townsend's law:

$$n(x) = n_0 \cdot \exp[\alpha x]$$

where $n(x)$ is the number of charges produced at a distance x from the origin, starting from $n_0$ primary charges; $\alpha$ is the first Townsend coefficient, the number of secondary charges formed per cm. The drifting charges produce current signals which are, for the electrons,

$$i_-(t) = \frac{n_0 e}{T_-} \exp[\alpha v_- t] \quad \text{for} \quad 0 < t \leq T_-$$

$$= 0 \qquad \qquad \text{for} \quad t > T_- \tag{20}$$

and for the positive ions,

$$i_+(t) = \frac{n_0 e}{T_+} \{\exp[\alpha v_- t] - \exp[\alpha v' t]\} \quad \text{for} \quad 0 < t \leq T_-$$

$$= \frac{n_0 e}{T_+} \{\exp[\alpha d] - \exp[\alpha v' t]\} \quad \text{for} \quad T_- < t < T_+ \tag{21}$$

where $1/v' = 1/v_- + 1/v_+$

$v_-$ and $v_+$ are the drift velocities of electrons and ions respectively;

$$T_- = \frac{d}{v_-} \quad \text{and} \quad T_+ = \frac{d}{v_+}$$

are the transit times over the interelectrode gap of a distance d; $n_0$ is the number of primary charges at the cathode (the grid in this case), and e is the elementary electron charge. The current signals are shown schematically in Fig. 15. The figure is not drawn to scale since

$$\frac{i_-(T_-)}{i_+(T_-)} \sim \frac{v_-}{v_+} \cong 1000 \quad \text{and} \quad T_- \cong 10 \text{ nsec}$$

$$\text{and} \quad T_+ \cong 5 \text{ μsec}.$$

In the case of the fine grid detector, the fast electron signal is used for the localisation of a photon event. The width of the exponential rise is equal to the transition time of the electrons across the gap (10 nsec).

Fig. (15) shows the positive ion component as measured on the fine grid electrode. As expected, there exists a plateau due to the constant drift velocity of the ions across the amplification gap. Some important characteristics of the detector can be derived from the signal shape.

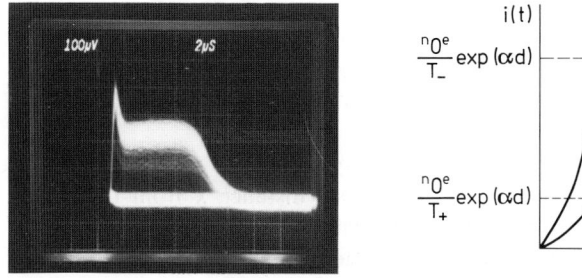

Fig. 15. Signal shape of a parallel electrode detector

## 5.3 Results

The height of the plateau in Fig. 15 is given by the value of the ion current at the moment T.

The last term in expression (3) can be neglected. From the result in Fig. (15) one calculates an amplification factor $\exp(\alpha d) = 2,5 \cdot 10^4$. This value is comparable with the gain achieved in classical multi-wire proportional chambers.

From Fig. (15) we conclude that the transit time for the slowly moving positive ions across the amplification gap of 1,5 mm width, is 5 μsec. This figure is given by the length of the plateau. This means that 5 μsec after the detection of a photon, the gap is cleared of any positive ions, so that no distortion of the electric field is possible anymore. It is known that the dimensions of clusters of primary electrons in argon, due to the absorption of a 8 keV X-ray is less than 0,5 mm in diameter [8]. As an estimate we can conclude that the count rate of this type of detector is at least $10^5 \text{ sec}^{-1} \cdot \text{mm}^{-2}$ before any degradation in amplitude by space charge effect takes place. It should be stressed that, as a consequence of the continuous nature of the electrode structure, the counting rate capability has to be calculated per unit area, and not per unit wire length, as in the case of discrete anode wires. This important conclusion has been confirmed by test results obtained with a similar device [34].

During the first tests with the fine grid device of Fig. 13 a spatial resolution of 0,5 mm FWHM was obtained, despite the rather poor quality of the delay-line.

The device of Fig. 14 is only a one-dimensional detector. For two-dimensional detectors, several schemes for the localization of avalanches are feasible. However, this needs a careful study of the signal distribution as is shown by the results obtained at CERN [34].

Parallel electrode devices are certainly very promising as PSD's for synchrotron radiation applications. The spatial resolution can be good in both, X- and Y-directions, because no quantization takes place along individual wires.

# 6 Photo-Electronic Imaging Devices

## 6.1 Vidicon-Tube Based Detectors:

It is evident that imaging systems, developed for commercial video applications are good candidates for applications in the X-ray field too.

Although the demands for quantitative imaging are quite different from commercial applications, it is shown that these devices are certainly very promising for X-ray measurements.

The advantage that the basic components are commercially available is however, weakened by the fact that the physicist does not have the possibility to tailor the system at will. As we will see, certain drawbacks have to be accepted as long as the industry is not prepared to develop special systems for X-ray applications, at affordable development costs.

The main components of a Vidicon tube based system are shown in Fig. 16.

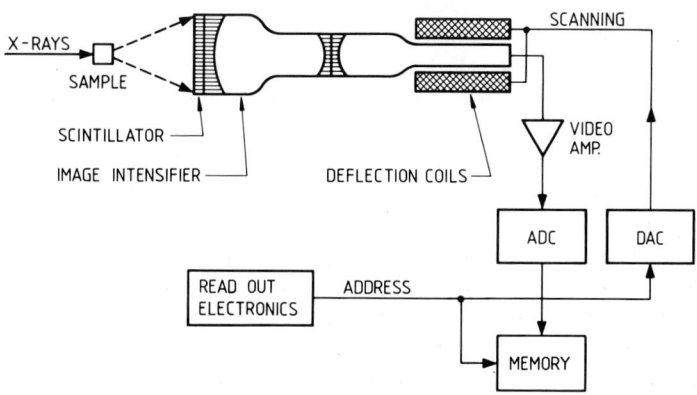

**Fig. 16.** Schematic of the Vidicon-tube detector

Since the photocathodes which are used in Vidicon tubes are sensitive to visible radiation only, some means of conversion of the X-ray photons in the keV-region to visible radiation has to be provided. This is achieved by installing a scintillator in front of the Vidicon.

Several kinds of scintillators are available, each of them with advantages and disadvantages. This component is crucial for the quality of the whole system as a detector. Intensive research is going on in this field, with the aim of providing good X-ray detectors for medical applications and for material testing.

A report of a detailed evaluation of some scintillators is given in [35]. The scintillator must have a high stopping power for the X-ray region of interest, to maximize the efficiency of the detector system. In addition, the phosphorescent light has to be matched to the spectral sensitivity curve of the photocathode. Usually an S-20 photocathode is installed, which has a maximum sensitivity at 420 nm. The most commonly used scintillator for X-ray applications is Zn S(Ag), which is settled from a suspension as fine grain polycrystalline layers. The same procedure is used for $Gd_2O_2S$. CsI(Tl) has some important advantages as a scintillator because it has a high stopping power and because it can be evaporated on a substrate in the form of a clear screen. There is quite a difference in thickness for 80% absorption of the CuKα-line, depending on the material used. The thicknesses are 160 μm for ZnS(Ag), 21 μm for $Gd_2O_2S$ and 12 μm for GsI(Tl) respectively [35].

This has important consequences concerning the spatial resolution of the detector system. Due to scattering and total reflection at the edges of the polycrystalline particles, more phosphorescent light is lost in a thick layer than in a thinner one. In addition, the visible photons are more spread-out laterally in thicker screens.

With modern Vidicon tubes the spatial resolution power of a detector system is predominantly determined by the optical properties of the scintillator.

An interesting development is the work done on CsI(Na) for medical radiology applications. CsI(Na) can be deposited on structured substrates in the form of fine pillars, which act as light guides, suppressing to a large extent the lateral spreading of the phosphorescent light [36]. A substantial improvement has been achieved in this way concerning the spatial resolution of X-ray detector systems. However, CsI-scintillators have the disadvantage that they are very sensitive to moisture. In fact, they have to be handled and stored in vacuum or in a protective gas atmosphere. Nevertheless, CsI-phosphors are the candidates for future systems.

The conversion of X-rays to photo-electrons from a photocathode is not very efficient. Careful measurements show that, in the average only 5 photoelectrons are released from an S-20 photocathode per 8 keV X-ray [35]. Inevitably this number is underlying statistical fluctuations, which has consequences for the amplitude accuracy of a vidicon system. Some amplification mechanism is necessary in order to produce usable signals from the few photoelectrons. One way is to use a separate image intensifier, which produces again visible photons on a phosphorescent screen. The light gain which can be achieved typically is as high as 250. Several stages of intensifiers can of course be cascaded. The disadvatage of such systems is some degree of loss in spatial resolution in the imaging phosphors.

Modern vidicon tubes with an integrated amplifying system exist in the form of the Silicon-Intensifier-Tanget-Tube (SIT-Tube). The main structure of this family of devices is shown in Fig. 17. It consists of an image-intensifier section, followed by a

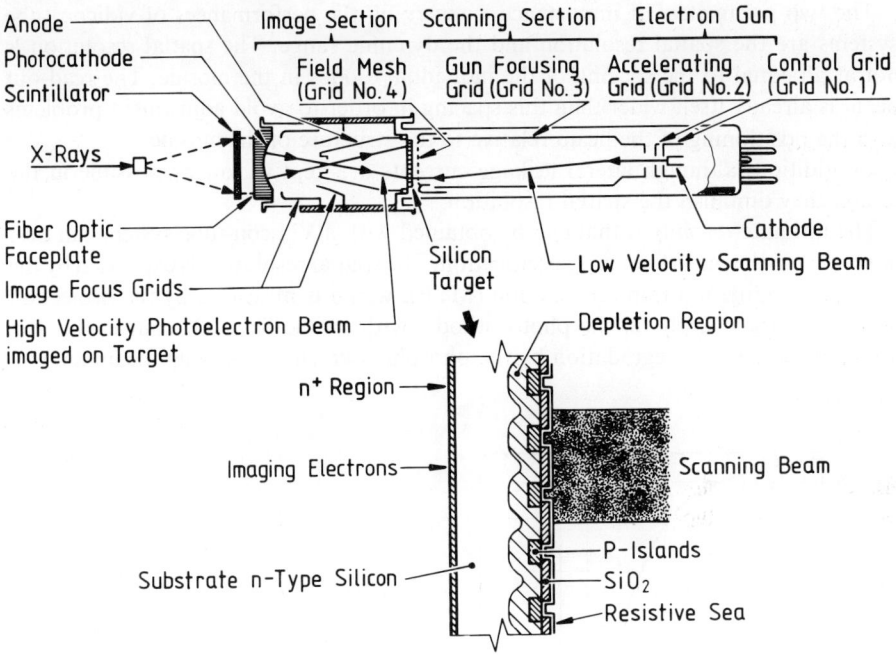

**Fig. 17.** Structure of an SIT-imaging tube, with principle of the target

silicon diode-array target. Betweeen the photocathode and the target, a high voltage is applied, which accelerates the photo-electrons liberated from the photocathode to such an extent that secondary electrons are produced in the target. The target is a matrix of separate semiconductor diodes implanted on a dice of silicon n-type material, some 10 μm thick (see Fig. 17). The density of the photodiodes is as high as 1800 elements per inch. The targets in use at present, are circular with up to 32 mm diameter. The front end of the target has an n-layer acting as a contact. The diodes are formed by p-diffusions on the back side forming islands, which are isolated from each other by $SiO_2$ deposits. The p-fields are typically 10 μm in diameter, on 20 μm centres. Accelerated electrons from the amplifying section of the SIT penetrate into the n-layer producing avalanches of secondary charges. The charge gain is dependent on the high voltage gradient, ranging from 200 to 2500. Each photo-electron released from the photocathode thus produces up to 2500 charge carriers in the target. The array is scanned by a focussed electron beam, produced by a thermoionic cathode and an electro-optic structure using focussing and deflection coils. The electron-beam deposits a uniform amount of electric charge on each of the diodes. The avalanches of secondary electrons discharge the diodes. When the diode charge is replenished again by the scanning beam, the current is a measure of the number of photoelectrons that have been deposited on the particular area. The target has a high resistivity, since it consists of reversely biased diodes, such that charges can be stored into it. In fact, each diode acts as a capacitor which is charged by the scanning beam and discharged by the photoelectrons.

The two parameters of importance, concerning the performances of vidicon tube systems are the spatial resolution and the dynamic range. The spatial resolution is not at all equal to the spacing of the individual diodes in the mosaic. The read-out beam is already itself wider than this spacing in order to avoid adjustment problems with the positioning of the beam relative to the structure of the mosaic.

In addition, although lateral leakage currents are kept as low as possible in the target, they diminish the spatial resolution.

The ultimate resolution that can be obtained with a Vidicon-tube system can best be seen from the manufacturers specification. The spatial resolution is expressed by the so called modulation-transfer function (MTF), which is measured by the projection of a test mask directly on the photocathode, with visible light. Of course, this test does not include the degradation by the phosphor screen in X-ray applications.

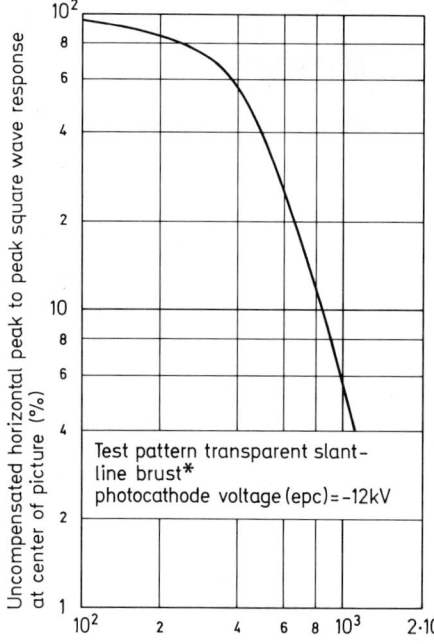

**Fig. 18.** Example of a Modulation Transfer Function of an SIT-tube

Fig. 18 reproduces the MTF curve of an 80 mm SIT-tube with a fiber optic faceplate (RCA (21146)). Plotted is the modulation, in precentage, of the video signal as a result of the projection of a test mask, showing bursts of dark-light bars with increasing spatial frequency.

The percentage modulation is plotted against the number of TV-lines per picture heigth. We can see that in order to measure really separated signals — corresponding to reflextions in X-ray work — the maximum number per picture height is 100 reflections. This corresponds roughly to a distance between reflections of at least 0.8 mm.

The specification for the resolution as given by manufactures as the maximum number of TV-lines is somewhat misleading. As seen in Fig. 18, this corresponds to a modulation of only 5%. This modulation is obtained at a distance between bars in the test pattern approximately equal to the full-width-half-maximum of the spatial response. In the center zone of an image this FWHM corresponds then to 80 μm, which is comparable with what can be obtained with gas filled detectors.

Bearing in mind the overall dimensions of the largest and most expensive Vidicon tubes with 80 mm photocathode diameter, the disadvantage in synchrotron radiation work becomes evident. The focal spot of the synchrotron radiation beam is much larger than 80 μm, and so are the reflections in a diffraction pattern, if the entire intensity of the source is to be used.

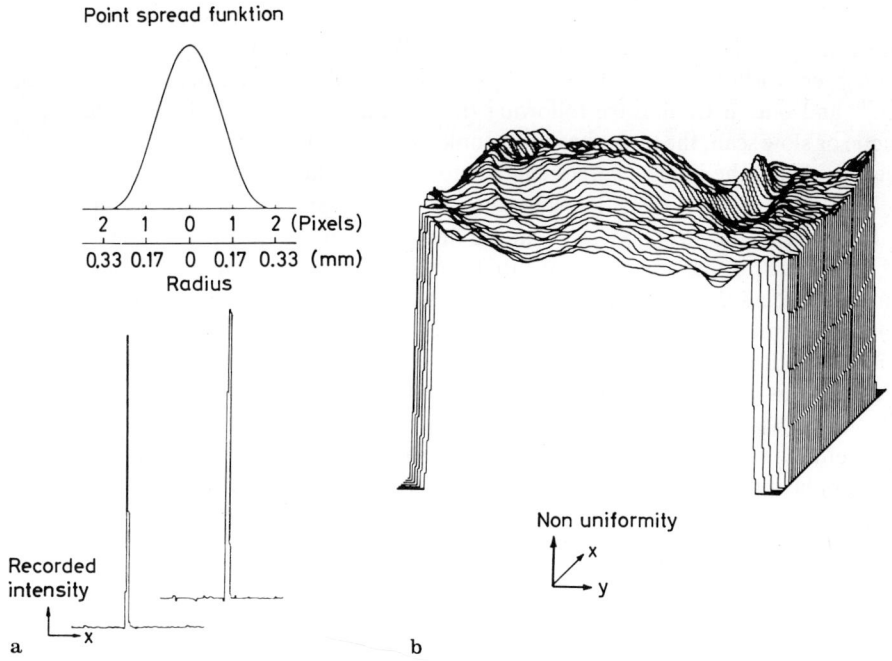

**Fig. 19a and b.** Performances of an SIT-tube based X-ray detector. **a** The response to an illumination through a 35 μm pinhole; **b** The sensitivity over the area of the detector

The performances of a Vidicon system as an X-ray detector are best demonstrated by Fig. 19 from [37]. Shown is the point-spread function as the response to an illumination of the detector, including a ZnS(Ag)-phosphor screen, through a 35 μm pinhole.

For this test, the target was scanned by the read-out beam in $256 \times 256$ steps. The full width of the point-spread function (PSF) is 3 pixels, corresponding to 43 "TV-line pairs". At FWHM, the width is 1.5 pixels, or 85 "TV-line pairs". Thus the 256 pixels in both directions are not at all independent. This result clearly shows the degradation of the imaging performances of a Vidicon system by the scintillation screen.

The sensitivity of a vidicon detector over its area is subject to nonuniformities of the scintillator screen, the photocathode, and the target. A typical result of a sensitivity test is shown in Fig. 19b from [37]. It is very stable and therefore the data can be corrected in a reliable way.

The dynamic range of vidicon systems is limited by the noise level of the video preamplifier on the low intensity side, and by the maximum read beam current at high local intensities. The gain of the amplificing stage has to be adjusted in order to optimize the dynamic range according to the requirements. The maximum read-out current results in a severe local intensity limitation of typically 5000 X-rays per pixel and per second. The maximum dynamic range obtained in actual systems is about 250:1 at best.

There are two modes of operation of TV-tube based detector systems: The integrating mode or "slow scan", and the TV-mode or "fast can". Two groups have been working intensively on TV-systems, one in Princeton using the first method [37, 38] and one in Cambridge following the second procedure [39]. In the integrating mode or slow scan, the target has to be cooled to low temperature, to minimize charge leakage. Then the image charges can be stored in the target and read-out later.

The gain of the intensifier stage has to be adjusted such that no saturation takes place at the strongest peaks in the image within the measuring time.

The slow scan has the advantage that the Video-amplifier bandwidth can be kept small, reducing the noise contribution of this element. Since a high number of events can be accumulated per pixel, a high resolution analog to digital converter is necessary to digitize the result. The data of several scans can be accumulated in a digital memory. Because this method relies on the measurement of the charge per pixel with high resolution, the read-out becomes necessarily slow, as a consequence of the conversion time of high resolution analog-to-digital converters. This means that in high counting rate applications, where the maximum number of counts per pixel, before saturation of the target elements, is collected in a short time, the duty cycle will be necessarily low. This can be a serious disadvantage in a synchrotron radiation experiment.

The TV-mode or "fast scan", uses the normal TV-scan rate of 50 pictures per second. Thus charges are accumulated over a period of 20 msec only and then sequentially digitized. In this mode, the probability that saturation occurs within a period of 20 msec is negligible, even with synchrotron radiation sources.

The bandwidth of the video amplifier has to be higher and its noise contribution, which is proportional to the square root of the bandwidth, will be more important. The ADC, to convert the number of charges into a digital number, has to be fast, but with lower resolution. Modern video-ADC's with up to 9 bit resolution and converion rates of 20 MHz are certainly helpful.

The accumulation and build-up of an image has to be done in a digital store. Here the demand of speed becomes quite high. Quantitative results of X-ray measurements with a "fast scan" system, are not available at present.

The Princeton Vidicon-tube system has been used successfully at EMBL as a detector for the high intensity reflection of a muscle diffraction pattern [38].

The system with a slow scan read-out is now routinely in use for SAXS-experiments on a rotating anode X-ray generator.

Extensive test have been carried out with a vidicon X-ray detector, based on a Westinghouse 80 mm SIT (TEM 432/R) at the DORIS-storage ring in Hamburg [40].

The data are at present recorded on video tape for later digitization and processing. The scintillator screen, deposited on a fiber optics face plate is either ZnS(Ag) or Gd S O(Tb). A spatial resolution of 0,5 mm FWHM has been measured. The system has been used in diffraction measurements of polymer samples and in protein crystallography as well. At present a quantitative analysis of the collected data is in progress.

As a conclusion, it can be said that TV-systems are not suited for time-resolved measurements with time frames as short as milliseconds. Even the "fast scan" is still much too slow to achieve an acceptable duty cycle with short time frames.

In addition, even with the largest photocathodes of 80 mm diameter, the focal spot of a synchrotron radiation source still has to be cut-down to enable the resolution of an appreciable number of reflections on the rather small detector area.

For small angle scattering work and for diffuse patterns, with high intensities as encountered in polymer physics applications, vidicon tube systems are certainly a possibility.

Although vidicon-tubes are commercial devices, it still takes a long way to turn them into an excellent X-ray detector. For accurate amplitude measurements, quite some effort has to be invested in the development and tests of the different components.

## 6.2 Charge Coupled Devices (CCD), and Photo Diode Arrays (PDA)

A new type of imaging devices has been introduced a decade ago [41]. These devices consist of an array of photo-sensitive diodes, coupled to a CCD read-out system. CCD's are based upon the transfer of charge packets in a repetitive metal-oxid°-silicon (MOS) structure (see Fig. 20). The charges in the elements of a CCD can be photo-generated, e.g. by the absorption of X-rays in the substrate, and then shifted out electronically to a single output. In a way CCD's operate as the solid state self-scanning equivalent of a vidicon-tube.

It has been shown that CCD's are directly sensitive to X-rays, in the range of 1 keV to 10 keV [42]. X-rays that penetrate the substrate create electron-hole pairs, which are trapped locally in so-called "potential wells", which are formed by appropriate biasing (see Fig. 20). Charges are thus stored in much the same way as in the cases of a vidicon target. By applying alternative voltages at subsequent time intervals, the potential wells, and the charges are shifted towards the single output of the device [43].

CCD's have found some limited applications as X-ray detectors [42,44].

The spatial resolution with CCD's is given simply by the dimensions of the individual cells, which are of the order of 10 μm to 20 μm. Linear devices, with up to 4096 elements are available from a number of manufactures. The major disadvantage, which makes CCD's almost useless in small-angle scattering and diffraction work, is their overall dimensions. Cell sizes are typically 13 μm × 17 μm, with overall lengths of up to 25 mm. Two-dimensional arrays are becoming available with as many as 500 × 500 picture elements. The industry adopted a standard conforming with the super 8 camera norm, which has a total picture size of 7 mm diagonally.

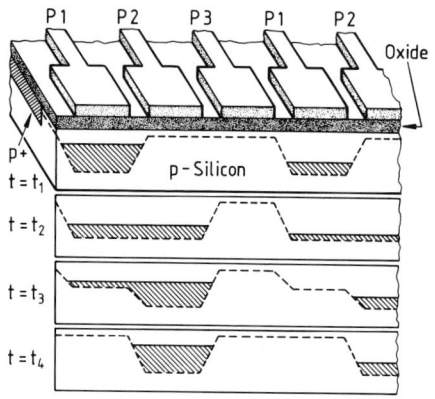

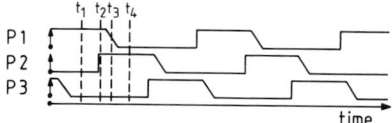

**Fig. 20.** Principle of a linear array Charge Coupled Device (CCD). (From Ref. [43])

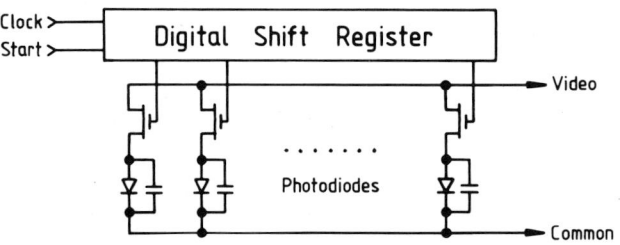

**Fig. 21.** Schematic of a linear Photo-Diode-Array (PDA)

Therefore, the CCD-type of imaging devices will be reserved to a few high resolution X-ray applications.

Another kind of linear solid state position sensitive detectors are the Photo-Diode-Arrays (PDA's), which are different from CCD's. A PDA consists of an array of separate photodiodes, each with an associated capacitance and a multiplexing read-out system (see Fig. 21). The charges collected in each cell are simply switched to the output, one by one. Unlike in the case of CCD's, the photosensitive elements are separated completely from the transfer circuity.

This has the consequence that the photodiodes can be made longer, which is of importance for X-ray applications. Cell dimensions of up to 25 µm × 2.5 mm are commercially available in arrays with up to 1024 elements [45]. The spatial resolution is given by the physical width of the individual cells, which in this case is 25 µm.

The dynamic range of PDA's is limited by the dark current in the photodiodes and by the maximum charge that can be stored in the cells. Dynamic ranges of up to

3000:1 have been reported, if the devices are cooled to liquid nitrogen temperatures [46]. At normal temperature, the dark current is far too high to permit any useful measurements.

Linear PDA's have found some application as X-ray detectors [46,47]. The overall dimensions make PDA's useful for some special applications, where a 25 µm resolution is required. In general, however, and specially with synchrotron radiation sources, PDA's do not make efficiently use of the available photon flux.

The largest two-dimensional PDA-device, which is commercially available, contains $128 \times 128$ elements, with overall dimensions of $8 \times 8$ mm. Again, it may find applications in some special cases, but not generally for X-ray detection.

## 7 Data Acquisition Systems

Electronic Position Sensitive Detectors, linear and area detectors, make the collection of the information contained in diffraction or scattering patterns highly efficient.

Patterns can be collected with meaningful statistics within fractions of seconds, if strong scatterers are studied, or they can be built up in repetitive measurements if reversible changes are studied. As a consequence, complete one-dimensional or two-dimensional image patterns can be collected in consecutive time intervals or time frames. The image information of each consecutive time frame has to be accumulated in separate memory segments. Thus time is added as another dimension to the one- or two- dimensional position information.

The main task of data acquisition systems for time resolved measurement is to record this three-dimensional data patterns. The amount of information which has to be recorded can become quite high.

The third development, which is essential to the successful application of position sensitive detectors in time resolved measurements is the availability of large and fast memory systems. As an example, a two dimensional detector with $256 \times 256$ pixels, in an experiment with 128 consecutive time frames of one second each will produce 8 M words of information within 128 seconds. In many small angle scattering experiments carried out at EMBL, time frames as short as 1 millisecond are no exception. This is the case for instance in studies of the diffraction patterns of contracting muscle [48], where up to a hundred repetitive measurements are accumulated. The total amount of data in such an experiment can be as high as $256 \times 256 \times 256 = 16$ Mwords.

The above considerations make clear that, unless an elaborate data acquisition system is available, it is advisable to reduce the amount of data. In small angle scattering experiments, for instance, where the information is one-dimensional, the use of a circular detector, reduces the large amount of data to a one-dimensional pattern.

Because of the counting rate capability of single photon counting PSD's, the access time of memory systems has to be short. Otherwise the counting rate limitation could well be determined by the data acquisition system. However, a large number of experiments do not require the ultimate speed. A data acquisition system with a one microsecond access time is perfectly feasible for many meaningful experiments, even with synchrotron radiation.

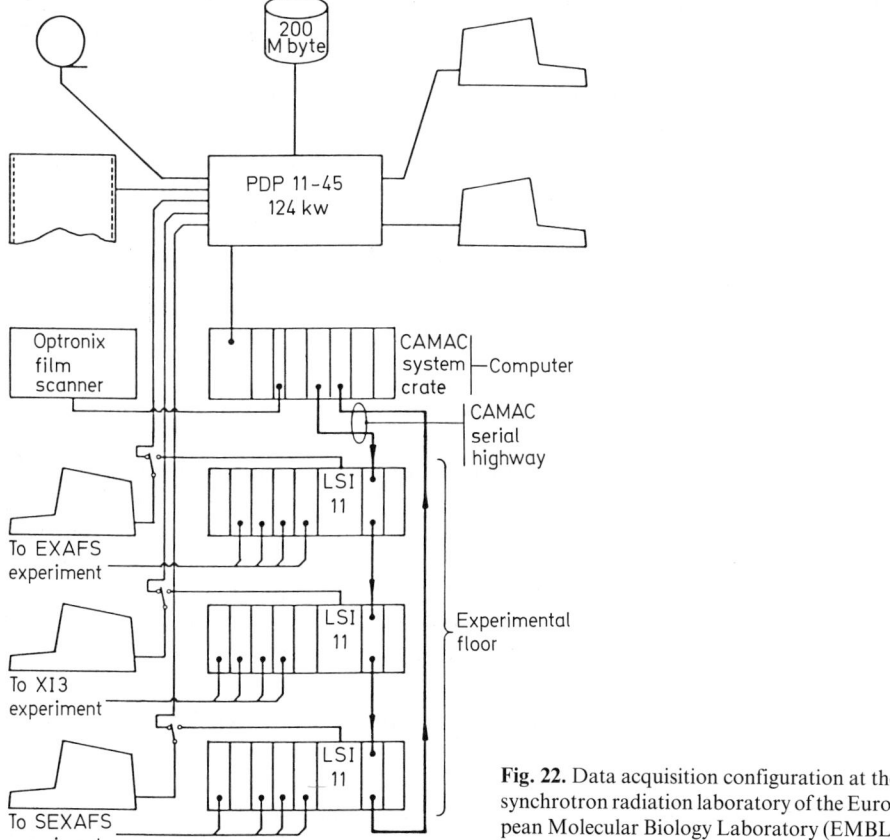

Fig. 22. Data acquisition configuration at the synchrotron radiation laboratory of the European Molecular Biology Laboratory (EMBL)

The diversity of the experiments with synchrotron radiation requires specialized electronic systems and a high degree of flexibility.

It is very helpful in a synchrotron radiation laboratory if a modular standard can be adopted. This situation is very similar to high energy physics where the CAMAC — standard has been developed [49]. This standard enables a high degree of modularity in both hardware and software.

For this reason the data acquisition systems at EMBL are based on CAMAC. Fig. (22) shows schematically the data acquisition configuration as it is used in our laboratory [50]. The high degree of modularity enables a continuous expansion of the system as required by the experiments, minimizing the development of new hardware and software.

Maybe the most important advantage of the CAMAC standard is the fact that a large number of modules with different functions are commercially available as a spin-off of high energy physics developments. As a result a substantial amount of designing and building time can be saved because only some adaptions in the standardized CAMAC software have to be made.

As an exmample of a data acquisition system for synchrotron radiation experiments, the set up of Fig. 22 will be discussed.

The main computing facility is a PDP-11-45 with 124 kW of main memory and 200 Mbytes of disk store. In addition good magnetic tape facilities for final storage of data are available. Presently the PDP-11-45 has become mainly the data acquisition computer and a VAX-750 has been installed for the data analysis. Each of the experiments has an LSI-11 in its first crate. This LSI-11, a microprocessor based crate controller, carries out the control of each experiment and the data acquisition sequences. In addition it formulates displays of collected data so that the experimenter can appraise the quality of the collected data before they are sent to the PDP-11-45. As an example of the set up of a time-resolved X-ray experiment with PSD, Fig. 23 shows a diagram of the data acquisition system [51].

The readout electronics in the schematic depend on the detector used and has already been discussed in Sect. 4.2. The sequencing of a time resoved measurement is carried out by the time-frame generator (TFG). This module is loaded with the desired sequence of time frames, not all of which have to have the same length. In addition the TFG is responsible for the synchronization of the measurement with the experiment (e.g. temperature control, stretching unit of the sample). The TFG divides the memory in pages for the different time frames by generating the frame addresses. The calibration channel unit, measures several parameters concerning the sample, for instance, the temperature, the stretching force, by pressure, appropriate transducers. In addition, the unit monitors the intensity of the primary X-ray beam, as the current of an ionisation chamber. This is important to normalize the measurements because the beam intensity of a synchrotron radiation source is not

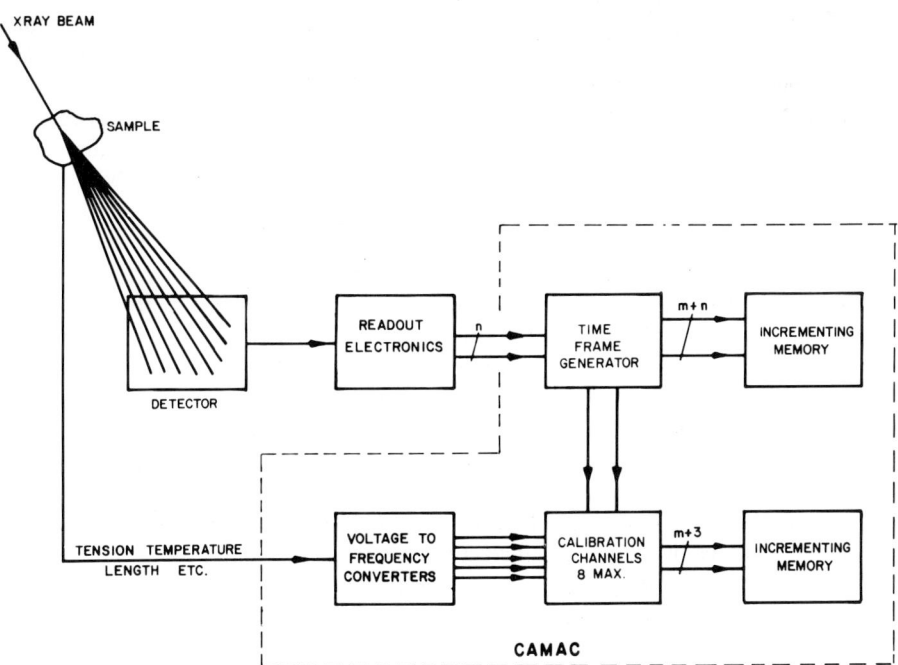

**Fig. 23.** Data acquisition system for time resolved measurements (EMBL-Hamburg)

constant, but normally decays between fillings of the storage ring. The self-incrementing memory is a 64 kWord CAMAC-module with a front input bus for addresses and the 16 bit data words. From the detector readout electronics a binary word, containing the address of the pixel is sent to the memory. In a handshake sequence the content of the addressed memory cell, is incremented by one in a read-modify-write cycle. Typically such a cycle takes 1 µsec.

With the advent of faster 2D-Detectors, and the enormous amount of data produced in certain experiments, it becomes necessary to have a system which is more powerful. One such system is in development at EMBL-Hamburg with the probably most challenging X-ray experiment in mind: X-ray crystallography. A high resolution 2D-Detector would produce such an amount of raw data within a short time period, that the available disk space would become a bottleneck. Therefore, a new system is now being developed around a bench of local processors, which can be programmed to do an on-line data reduction, before they are sent to the host computer. Details about the new system can be found in Ref. [52].

# 8 Conclusion

The developments which have been described in this review show that many groups are working in this field, along different lines and for a variety of applications. Each method, be it gas filled single photon counting counters or photoelectronic imaging devices has its own merits for a particular experiment.

There will probably never exist a kind of universal detector for all applications. New ideas and new technological possibilities, especially in the field of electronics will bring innovations in the detector development and hence, new prospects for more sophisticated experiments.

Gas filled detectors, if they are well designed are not old-fashioned devices, but are highly developed pieces of equipment, which allow very accurate measurements. They can be tailored and optimized for a variety of applications, because the physicist has access to their principle of operation. In addition, they can be made large enough, such that full use can be made of the intensity of a synchrotron radiation source, with the present focal spot dimensions. The relative and even the absolute spatial resolution is very good. There is no need for a 10 micrometer resolution in most applications but more for 200 micrometer.

Photoelectronic devices are very promising for a number of experiments. Certainly they have the advantage that they are integrating devices. Unless great care has been taken in the design they are not yet quantum limited devices.

The counting rates encountered in synchrotron radiation experiments may be high, but not that extreme that single photon counting devices are hopeless. The quality of the data that are recorded, within a given amount of time is what ultimately dictates the choice of a detector system.

The newer developments concerning gas filled detectors based on parallel electrode structures and the associated electonics are very promising concerning spatial resolution and counting rate capability.

The parallel electrode structure with no very fine wires and with appropriate gas fillings, will turn gas filled detectors into very reliable devices.

May be in the future, the industry will develop some kind of large CCD devices, with faster readout systems. This really would bring an innovation. In the meantime it is worth while considering that gas filled detectors already serve their purpose very well at present.

# 9 Appendix

## The Derivation of the Expression for the Gas Amplification

Electrons, generated by the photoelectric effect, upon the absorption of an X-ray photon, are swept towards the positively charged anode (see Fig. 1). If E(r) is sufficiently high, the drifting electrons acquire enough kinetic energy to ionize other neutral gas atoms by collision. Thus, secondary charges are created in the form of an avalanche. This is the basis of gas amplification.

As already mentioned in Section II.A, the increase in the number of electrons, over a path dx in the electric field, is given by

$$dn = n_0 \alpha \, dx \quad (A.1)$$

where $\alpha$ is the first Townsend coefficient. $\alpha$ is a function of the local electric field, and the expression for the multiplication of the number of charges, obtained by integration of (A.1) takes the general form:

$$A = \frac{n}{n_0} = \exp\left[\int_{x_1}^{x_2} \alpha(x) \, dx\right] \quad (A.2)$$

In a cylindrical proportional counter, the electric field has a radial distribution as shown in Fig. 1a. The field is a function of r, the radial distance from the center, and is given by

$$E(r) = \frac{V_b}{\ln\left(\frac{b}{a}\right)} \cdot \frac{1}{r} = \frac{V(r)}{\ln\left(\frac{r}{a}\right)} \cdot \frac{1}{r} \quad (A.3)$$

where a and b are the radii of the anode and the cathode respectively. The magnitude of the field is shown schematically in Fig. 1b.

Diethorn [2] derived an expression for the gas amplification in a coaxial chamber system starting from the following assumptions.

It is postulated that $\alpha$, the first Townsend coefficient, is a function of E/P, the ratio of the electronic field in V/cm over the pressure of the filling gas in Torr, and that,

$$\alpha = 0 \quad \text{for} \quad E/p < K$$

$$\alpha \sim E/p \quad \text{for} \quad E/p > K \quad (A.4)$$

$$\alpha = \alpha_0 \quad \text{for} \quad E/p = K$$

It follows that there exist a critical field strength, $E_c$, below which no gas amplification takes place. From (A.4) it follows that

$$E_c = Kp$$

Corresponding to E, there exist a critical radius $r_c$, determining the critical volume around the anode wire, within which avalanche formation can take place. At the distance $r_c$ from the centre, the potential is $V_c$. The mean free path, between ionization is, by definition

$$\lambda = 1/\alpha \tag{A.6}$$

At constant $\alpha$, $\lambda$ varies with $1/P$, the inverse of the gas pressure. On the other hand, with p kept constant, varies as $1/\alpha$. Accordingly,

$$\lambda \sim \frac{1}{\alpha p} \sim \frac{1}{\left(\frac{E}{p}\right) \cdot p} \sim \frac{1}{E} \tag{A.7}$$

Since, $\lambda$ is small, E is practically constant over $\lambda$. The work done on an electron, or the energy gained by it is given by

$$\Delta Ve = Ee \cdot \lambda \tag{A.8}$$

It follows that, under the above assumption, $\Delta V$, the potential difference crossed by the electron between ionizations is not a function of E and P. This means that $\Delta V$ is, like K, a characteristic constant of the gas.

Within the volume defined by $r_c$, where the potential is $V_c$ there will be

$$n = \frac{V_c}{\Delta V} \tag{A.9}$$

ion pairs produced.

Each ionization produces another electron, such that the gas multiplication is given by

$$A = 2^n \tag{A.10}$$

or

$$\ln A = n \ln 2$$

From (A.3) and (A.5) we can write

$$r_c = \frac{V_b}{E_c \cdot \ln\left(\frac{b}{a}\right)} = \frac{V_b}{Kp \ln\left(\frac{b}{a}\right)} \tag{A.11}$$

From (A.3) and (A.11) we obtain

$$V_c = \frac{V_b \cdot \ln\left(\frac{r_c}{a}\right)}{\ln\left(\frac{b}{a}\right)} \qquad (A.12)$$

combining (A.11) and (A.12) yields

$$V_c = V_b \cdot \ln\left[\frac{V_b}{Kpa \ln\left(\frac{b}{a}\right)}\right] \cdot \frac{1}{\ln\left(\frac{b}{a}\right)} \qquad (A.13)$$

which gives $V_c$ as a function of $V_b$, a, b and p. Substituting in (A.9) and (A.10), the expression for the gas amplification becomes:

$$\ln A = \frac{\ln_2}{\Delta V \ln\left(\frac{b}{a}\right)} \cdot V_b \cdot \ln\left[\frac{V_b}{Kpa \ln\left(\frac{b}{a}\right)}\right] \qquad (A.14)$$

Where $\Delta V$ and K are constants, specific for the gas filling of the detector.

## 10 References

1. Lang, A. R.: Rev. Sci. Instr., Vol. 33, 96 (1956)
2. Diethorn, W.: U.S. AEC Report NYO-6628 (1956)
3. Hendricks, R. W.: Nucl. Instr. and Meth., Vol. 102, 309 (1972)
4. Hendricks, R. W.: Rev. Sci. Instr., Vol. 40, 1216 (1969)
5. Walenta, A. H.: Nucl. Instr. and Meth., Vol. 217, 65 (1983)
6. Wilkinson, D. H.: "Ionization Chambers and Counters", p. 181, Cambridge: Cambridge University Press 1950.
7. Neuert, H.: „Kernphysikalische Messverfahren", p. 41, Karlsruhe: Verlag G. Braun 1966.
8. Sauli, F.: CERN Report, CERN 77-09 (1977)
9. Baldinger, E., Franzen, W.: Advances in Electronics and Electron Physics, Vol. 3, 256 (1956)
10. Radeka, V., Karlovac, N.: Nucl. Instr. and Meth., Vol. 52, 86 (1967)
11. Konrad, M.: IEEE Trans. Nucl. Sci., Vol. NS-15, No. 1, 268 (1968)
12. Radeka, V.: IEEE Trans. Nucl. Sci., Vol. NS-21, No. 1, 51 (1974)
13. Radeka, V.: Short Course on "Radiation Detection" IEEE Nuclear Science Symp. Oct. 1983 (see also Ref. 17)
14. Gatti, E., Longoni, A.: Nucl. Instr. and Meth., Vol. 163, 83 (1979)
15. Grove, R., Lee, K., Perez-Mendez, V. and Sperinde, J.: Nucl. Instr. and Meth., Vol. 89, 257 (1970)
16. Gabriel, A., Dauvergne, F. and Rosenbaum, G.: Nucl. Inst. and Meth., Vol. 152, 191 (1978)
17. Boie, R. A., Fischer, J., Inagaki, Y., Merritt, F. C., Radeka, V., Rogers, L. C., Xi, D. M.: Nucl. Inst. and Meth., Vol. 201, 93 (1972)
18. Allemand, R., Thomas, G.: Nucl. Instr. and Meth., Vol. 137, 141 (1976)
19. Borkowski, C. J., Kopp, M. K.: Rev. Sci. Instr., Vol. 40, 951 (1975)

20. Ford, J. L. C.: Nucl. Instr. and Meth., Vol. 162, 277 (1979)
21. Kopp, M. K.: Rev. Sci. Instr., Vol. 48, 383 (1977)
22. Charpak, G., Bouclier, R., Bressani, J., Facrier, J., Zupancro, C.: Nucl. Inst. and Meth., Vol. 62 235 (1968)
23. Erskine, G. A.: Nucl. Instr. and Meth., Vol. 105, 565 (1972)
24. Hendrix, J.: EMBL Research Report 1982
25. Gabriel, A., Dauvergne, F.: Nucl. Inst. and Meth., Vol. 201, 223 (1982)
26. Hamlin, R.: Transaction ACA, Vol. 18, 95 (1982)
27. Helliwell, J. R., Hughes, G., Pryzbylski, M. M., Ridley, P. A., Sumner, I., Bateman, J. E., Connolly, J. F., Stephenson, R.: Nucl. Instr. and Meth., Vol. 201, 175 (1982)
28. Radeka, V., Boie, R. A.: Nucl. Instr. and Meth., Vol. 178, 543 (1980)
29. Boie, R. A., Fischer, J., Inogaki, Y., Merrit, F. C., Okuno, H., Radeka, V.: Brookhaven Nat. Lab. Report BNC-30650 (1981)
30. Parkman, C., Hajduk, Z., Jeavons, A., Ford, N., Lindberg, B.: Proc. 2nd ISPRA Nuclear Science Symposium EUR 530e, 237 (1975)
31. Kahn, R., Fourme, R., Bosshard, R., Candron, B., Santiard, J. C., Charpak, G.: Nucl. Inst. and Meth., Vol. 201, 203 (1982)
32. Charpak, G., Demierre, C., Kahn, R., Santiard, J. C., Sauli, F.: Nucl. Inst. and Meth., Vol. 141, 141 (1977)
33. Hendrix, J.: IEEE Trans. Nucl. Sci., Vol. NS-31, No. 1, 281 (1984)
34. Peisert, A.: Nucl. Instr. and Meth., Vol. 217, 229 (1983)
35. Gruner, S. M., Milch, J. R., Reynolds, G. T.: Princeton Univ., Dept. of Physics, Techn. Rep. No. 7, (1983)
36. Stevels, A. L. N., Schrama, A. D. M.: Philips. Res. Rep., Vol. 29, 340 (1974)
37. Gruner, S. M., Milch, J. R., Reynolds, G. T.: Rev. Sci. Instr., Vol. 53, 1770 (1982)
38. Milch, J. R., Gruner, S. M., Reynolds, G. T.: Nucl. Instr. and Meth., Vol. 201, 43 (1982)
39. Arndt, U. W., Gilmore, D. J.: J. Appl. Cryst., Vol. 11, 113 (1978) and Adv. Electr. Elect. Phys., Vol. 52, 209 (1979)
40. Prieske, W., Riekel, C., Zachmann, H. G.: Nucl. Instr. and Meth., Vol. 208, 435 (1983)
41. Boyle, W. S., Smith, G. E.: Bell Syst. Techn. J., Vol. 49, April (1970)
42. Catura, R. C., Smithson, R. C.: Rev. Sci. Instr., Vol. 50, 219 (1970)
43. Sequin, S. H., Tompsett, M. F.: "Charge Transfer Devices", Adv. Electr. Elect. Phys., Suppl. 8 (1975) 44. Koppel, L. N.: Rev. Sci. Instr., Vol. 48, 669 (1977)
45. Reticon, Data Sheet, S-Series Solid State Line Scanners.
46. Borso, C. S., Danyluk, S. S.: Rev. Sci. Instr., Vol. 1669 (1980)
47. Borso, C. S.: Nucl. Instr. and Meth., Vol. 201, 65 (1982)
48. Huxley, H. E.: Nucl. Instr. and Meth., Vol. 201, 123 (1982)
49. CAMAC Instrumentation and Interface Standards, IEEE Publication, Wiley Interscience (1976)
50. Clout, P. N.: Nucl. Instr. and Meth., Vol. 201, 225 (1982)
51. Boulin, C., Dainton, D., Dorrington, E., Elsner, G., Gabriel, A., Bordas, J., Koch, M. H. J.: Nucl. Instr. and Meth., Vol. 201, 209 (1982)
52. Nicolae, G.: Proc. of the SPIE., Vol. 435, 173 (1983)


H.-G. Zachmann (Editor)
Received July 7, 1984

# Fluorescence Anisotropy Technique Using Synchrotron Radiation as a Powerful Means for Studying the Orientation Correlation Functions of Polymer Chains

Jean-Louis Viovy and Lucien Monnerie
Laboratoire de Physicochimie Structurale et Macromoléculaire, Ecole Supérieure de Physique et de Chimie 10, rue Vauquelin 75231 PARIS Cedex 05 (France)

*This article shows how the recent developments in the Fluorescence Anisotropy Decay technique permitted by Synchrotron Radiation can be used to study local polymer dynamics both in dilute solutions and in bulk polymers.*

*In section 2, the different theoretical models for local dynamics are briefly reviewed, and their connection with spectroscopic experiments is recalled. The Fluorescence Anisotropy Decay technique and the synchrotron source are presented in section 3. The fourth section is concerned with two typical examples. Using a series of experiments performed on polystyrene dilute solutions and another one performed on melt polybutadiene, we show how the different theoretical models can be told apart, and we present new information about the processes responsible of backbone rearrangement which has been obtained using the cyclosynchrotron LURE-ACO at Orsay (France).*

1  Introduction . . . . . . . . . . . . . . . . . . . . . . . 100

2  The Molecular Approach to Local Polymer Dynamics . . . . . . . . . . . . 100
   2.1 Spectroscopic Experiments and Correlation Functions . . . . . . . . 100
   2.2 Models for the Orientation Autocorrelation Function of an Isolated Chain  102
   2.3 The Case of Bulk Polymers . . . . . . . . . . . . . . . 104

3  Recent Progress in the Fluorescence Anisotropy Decay Technique . . . . . . 105
   3.1 Principle . . . . . . . . . . . . . . . . . . . . . . 105
   3.2 The FAD Apparatus . . . . . . . . . . . . . . . . . . 106
   3.3 The Synchrotron Radiation . . . . . . . . . . . . . . . 107
   3.4 Quantitative Criteria for Data Interpretation . . . . . . . . . 109
   3.5 Fluorescent Labeling of Polymer Chains . . . . . . . . . . . 109

4  Use of Fluorescence Anisotropy Decay for Studying the OACF of Polymers . . . 110
   4.1 Dilute Solutions of Labelled Chains . . . . . . . . . . . . 110
   4.2 OACF of Labelled Chains in a Bulk Polymer . . . . . . . . . 114
   4.3 Study of Temperature Effects . . . . . . . . . . . . . . 117

5  Conclusions . . . . . . . . . . . . . . . . . . . . . . . 120

6  References . . . . . . . . . . . . . . . . . . . . . . . 121

# 1 Introduction

The dynamic behavior of bulk polymers is of considerable practial importance, and it has found a long standing and wide-spread interest [1,2]. The first tools of these studies were mainly dielectric [3,4] and mechanical [2] relaxations. These techniques brought out several striking phenomenological properties of polymeric materials, such as the non-Arrhénius slowing down of dynamics when the temperature is lowered, the occurrence of a glass-transition and of secondary relaxations and the wide distribution of relaxation times in the spectra. However, these experiments involve cross-correlations between all the chains in the medium, and their molecular interpretation is difficult. Maybe due to this lack of experimental knowledge on a molecular scale, the theories for dynamics in polymer melts also remained on a rather phenomenological level [5-11]. Such a situation may evolve rapidly in the forecoming years.

In the case of dilute solutions of polymers, numerous studies using spectroscopic methods able to probe motions at a molecular level, such as NMR, ESR or Fluorescence Anisotropy Decay (FAD) revealed some molecular aspects of the dynamics, in particular the influence of chain connectivity on local processes [12]. This knowledge was recently improved thanks to the application of Synchrotron Radiation to FAD [13]; This technique provides a quantitative tool for discussing the numerous theoretical models proposed to account for single chain polymer dynamics, as shown in the following.

The understanding of bulk polymer dynamics involves new and difficult questions such as the relative importance of intra and inter chain constraints, or the relation between molecular motions and the complicated mechanical behavior of these materials. Numerous experiments on bulk polymers using NMR [14-17], ESR [18] or Fluorescence Polarization under continuous excitation [19,20] are presently developed. But these experiments need the a priori choice of a model of motion to be interpreted, and such models do not exist so far for the local dynamics in bulk polymers. This limitation is very troublesome, since experiments carried out on polymers in solution have shown that varying the choice of the model used in data treatment could lead to important discrepancies in the derived correlation times or activation energies. In the following, we will show how Fluorescence Anisotropy Decay may help to overcome this difficulty, and we will give some examples of original information that can be obtained using this technique in conjunction with the powerful synchrotron light source.

# 2 The Molecular Approach to Local Polymer Dynamics

## 2.1 Spectroscopic Experiments and Correlation Functions

An "exact" knowledge of the dynamics of a polymer chain would imply that the position $\underline{r}_i$ and the orientation $\underline{u}_i$ of each $i^{th}$ rigid part (bond, group ...) of the chain, and the time derivative of these quantities be known at any time. Such a knowledge is unattainable and, fortunately, not necessary.

Since molecular motions are stochastic, the only significant knowledge is statistical, and all the properties of a medium can be predicted from a rather low number of mean quantities depending on $\underline{r}_i$ and $\underline{u}_i$, and called correlation functions (CF). In particular, the results of spectroscopic experiments often can be expressed in terms of these C F, leading to a description of molecular dynamics which can be used for discussing theoretical models and predicting the macroscopic behavior under various circumstances.

The general form of a C F is

$$f(t) = \langle x(\underline{r}_i, \underline{\dot{r}}_i, \underline{u}_i, \underline{\dot{u}}_i, 0) \cdot y(\underline{r}_j, \underline{\dot{r}}_j, \underline{u}_j, \underline{\dot{u}}_j, t)\rangle$$

where x and y may be any scalar function of position and orientation, and the brackets represent the integral over phase space.

Since we are dealing with motions of the order of one bond length or more in a dense medium, inertial effects can be neglected, and the dynamics on that scale only involve correlation functions over $\underline{r}_i$ and $\underline{u}_i$. Different spectroscopic techniques may probe different C F. For instance, Neutron or polarized light scattering probe translation C F, (acting only or $\underline{r}_i$) such as the well known Van Hove C F:

$$G(\underline{r}, t) = \langle n_i(\underline{r}_0, 0) \cdot n_j(\underline{r}_0 + \underline{r}, t)\rangle_{ij} \qquad (1)$$

where $n_j(\underline{r}, t)$ is the number density, i.e. the probability that particle j is at position $\underline{r}$ at time t.

Depolarized Dynamic Light Scattering and Dielectric Relaxation are related to the the multimolecular orientation correlation function

$$F^{multi}(\underline{u}_0, \underline{u}, t) = \langle d_i(\underline{u}_0, 0) \cdot d_j(\underline{u}, t)\rangle \qquad (2)$$

where $d_j(\underline{u}, t)$ is the probability that particle j has orientation $\underline{u}$ at time t) while ESR, NMR, the Transient Kerr Effect, Fluorescence Polarization under Continuous Excitation and Fluorescence Anisotropy Decay (FAD) probe Orientation Auto Correlation Functions (OACF) related to:

$$F^{mono}(\underline{u}_0, \underline{u}, t) = \langle d_i(\underline{u}_0, 0) \cdot d_i(\underline{u}, t)\rangle \qquad (3)$$

The orientation of bonds is strongly affected by local molecular motions, and orientation CF reflect local dynamics in a very sensitive way. However, the interpretation of multimolecular orientation CF requires the knowledge of dynamic and static correlations between particles. Even in simple liquids this problem is not completely elucidated. In the case of polymers, the situation is even more difficult since "particles" i and j, which are monomers or parts of monomers may belong to the same chain or to different chains. Thus, we believe that the molecular interpretation of monomolecular orientation experiments in polymer melts is easier, at least in the present early stage of study. Experimentally, the OACF never appears as the complicated nonseparated function of time and orientation given in expression (3), but only as correlation functions of spherical harmonics

$$F^{mono}_{lm}(t) = \langle Y^m_l[\underline{u}_i(t)]\, Y^m_l[\underline{u}_i(0)]\rangle \qquad (4)$$

From the Curie principle and symmetry considerations it can be shown that for an isotropic medium, the first and second moments $F_{1,0}^{mono}(t)$ and $F_{2,0}^{mono}$ are generally sufficient for a complete description of the system [21]. The interactions used in spectroscopic experiments as ESR, NMR and FAD are affected only by the second moment of the OACF:

$$M_2(t) = F_{2,0}^{mono}(t) = \frac{1}{2} \langle 3 \cos^2 [\theta(t)] - 1 \rangle \tag{5}$$

where $\theta(t)$ is the rotational angle of $u_i$ during time t. ($\theta(t) = (\underline{u}_i(0), \underline{u}_i(t))$)

## 2.2 Models for the Orientation Autocorrelation Function of an Isolated Chain

It has been recognized for a long time that the orientation dependence of a vector fixed to a polymer chain could not be represented by a simple isotropic rotational diffusion model. In such a model [22] the orientation is assumed to follow a vector joining the center of a sphere to a point performing a random brownien diffusion on the surface of that sphere. According to this model which describes well the orientation of spherical objects or infinitely thin rigid rods, the OACF is an exponential function [23].

Generalized rotational diffusion models were built up later to account for the orientational relaxation of anisotropic rigid bodies [24–27] or for the orientational relaxation in an anisotropic medium [28,29]. But these models are very far from the actual flexible polymer chain, and one cannot expect to understand polymer dynamics without turning to models which take into account the molecular nature of polymers. In that case, each bond is subjected to particular anisotropic constraints due to neighboring bonds. Rouse [30] proposed to model the chain by a sequence of beads separated by springs. The random forces exerted by the viscous environment are localized on the beads. In spite of its crudeness, this early model contains the two essential features of polymer dynamics, i.e. the connectivity and the flexibility. It leads to a master equation for the orientation probability:

$$\frac{dP_\theta(x, t)}{dt} = W \frac{d^2P_\theta(x, t)}{d^2x} \tag{6}$$

Where $P_\theta(x, t)$ is the probability that a bond with curvilinear coordinate x along the chain has orientation $\theta$ at time t.

This Equation is of the one-dimension-diffusion type and it leads to a $t^{-1/2}$ long time dependence of the OACF [31]. This model is supposed to be valid only on a distance range greater than the "statistical unit", i.e. the smallest chain portion large enough to be gaussian. Thus, more realistic descriptions of the chain have been proposed, in order to get OACF expressions valid in the whole time and distance range of experiments.

In the model proposed by Valeur, Jarry, Geny and Monnerie (VJGM), the chain is assumed to perform 3-bond motions on a tetrahedral lattice, to account for the fixed bond angles imposed to real backbones [32]. This model leads to a one-dimension diffusion equation for the orientation probability, with the following expression for the OACF:

$$M_2(t) = \exp(-t/\tau_2) \exp(t/\tau_1) \operatorname{erfc}[(t/\tau_1)^{1/2}]$$

$\tau_1$ is related to the inverse of the rate of 3-bond motions, and the $\tau_2$-exponential term is introduced to account for out-of-lattice motions.

Relation (7) presents an unrealistic infinite first derivative at $t = 0$, due to the continuous approximation made in the analytical treatment, and further refinements were proposed to avoid this defect. For example, under the continuous approximation, Bendler and Yaris have performed an arbitrary truncation in the mode analysis [33].

The expression for the OACF is then:

$$M_5(t) = \frac{1}{2}\left(\frac{\pi}{2}\right)^{1/2}(\tau_1^{-1/2} - \tau_2^{-1/2})\{\operatorname{erfc}[(t/\tau_2)^{1/2}] - \operatorname{erfc}[(t/\tau_1)^{1/2}]\} \qquad (8)$$

This expression presents a satisfactory behavior at short time. However, it seems difficult to correlate the parameters $\tau_1$ and $\tau_2$, which are the inverses of the arbitrary cutoff frequencies, with molecular quantities.

Another solution is to use the master equation in its discrete from and to perform an exact mode analysis on the resulting Hückel matrix arbitrarily truncated. In such a case the truncation is directly associated with the finite length of the chain which is taken into account in the calculation. In fact, this procedure, proposed by Jones and Stockmayer [34] does not lead to a closed expression for the OACF, but to an infinite series of expressions corresponding to different truncations. This makes the comparison of the J S model with experiments rather lengthy. Since similar ideas can now be accounted for by closed expressions, we will not present here the detailed discussion of the JS model (for a more complete discussion, see Ref. [13]).

In contrast with these models, which start from a rather crude representation of the chain, but allow a direct computation of the OACF, Hall and Helfand [35] proposed recently a model able to predict conformational correlation functions (CCF) for rather realistic molecular potentials. The OACF cannot be derived from these CCF, at least at the present time. However, Hall and Helfand suggested that the CCF for a chain of two-state elements:

$$C_{ii}(t) = \exp(-t/\tau_2) \exp(-t/\tau_1) I_0(t/\tau_1) \qquad (9)$$

may be a good approximation for the OACF ($I_0$ represents the modified Bessel function of order 0). In that case, $\tau_2$ should be associated with isolated conformational jumps, and $\tau_1$ to correlated ones.

Several refinements of the HH model have been proposed recently [13,36,37]. For the sake of simplicity, we do not present here these models, nor their detailed discussion using FAD, since they rely on the same essential physical assumptions, and lead to the same physical conclusions as the original HH model. For a complete discussion, the reader is invited to refer to Ref. [13].

The different models presented briefly hereabove are not based exactly on the same representation of the polymer chain, and do not involve the same analytical treatment. However, it is worth to notice that all of them agree with the prediction of an OACF which contains the two following features (see Table 1).

**Table 1.** Expressions for the anisotropy time dependence used through the paper

| Abbreviation | Expression | Ref. |
|---|---|---|
| WW | $r(t) = r_0 \exp((-t/\tau_1)^\beta)$ | 48 |
| VJGM | $r(t) = r_0 \exp\left(\left(\dfrac{1}{\tau_1} - \dfrac{1}{\tau_2}\right)t\right) \mathrm{erfc}\,((t/\tau_1)^{1/2})$ | 32 |
| BY | $r(t) = 0.5 r_0 (\pi/t)^{1/2} (\tau_2^{-1/2} - \tau_1^{-1/2}) \cdot (\mathrm{erfc}((t/\tau_1)^{1/2}) - \mathrm{erfc}((t/\tau_2)^{1/2}))$ | 33 |
| HH | $r(t) = r_0 \exp(-t/\tau_2) \exp(-t/\tau_1) I_0(t/\tau_1)$ | 35 |
| IR | $r(t) = r_0 \exp(-t/\tau_1)$ | 23 |
| RR | $r(t) = r_0((1 - \beta) \exp(-t/\tau_1) + \beta)$ | 28 |

(i) A non exponential short time term, characteristic of a one-dimensions diffusion. This term, with characteristic time $\tau_1$, is associated with some "elementary motion" which must diffuse along the chain because of the connectivity.

(ii) An exponential loss, with characteristic time $\tau_2 > \tau_1$, which reflects some finite damping or truncation of this diffusion. The complex shape of the OACF of polymers in solution has already been observed experimentally [12,33,38–41] but, until the recent application of FAD under synchrotron excitation to polymer dynamics [13], the different models had not been held apart by experiments.

## 2.3 The Case of Bulk Polymers

The dynamics of bulk polymers have been approached in two different ways. On one hand, models of localized conformational jumps have been proposed to interpret numerous NMR experiments (see e.g. Ref. [17] or [42]). These models, which are specific of a given polymer assume that a short chain sequence performs conformational jumps between a few number of sites, the rest of the chain being immobile. Such localized jumps would lead to a well separated elastic peak in neutron quasielastic scattering experiments, in contradiction with all the experimental data obtained from polymer melts [43–47]. Indeed, these models can in some cases be invoked to describe secondary relaxations in glassy polymers, but they are not sufficient to account for the numerous liquid-like properties of polymer melts.

On the other hand, some phenomenological distributions of relaxation times, such as the well known Williams-Watts distribution [48] (see Table 1, WW) provided a rather good description of dielectric relaxation experiments in polymer melts, but they are not of considerable help in understanding molecular phenomena since they are not associated with a molecular model. In the same way, the glass transition theories account well for macroscopic properties such as viscosity, but they are based on general thermodynamic concepts as the free volume [7,10] or the configurational entropy [8,9], and they completely ignore the nature of molecular motions.

Very simple and fundamental questions are still without answer, and, in the following, we will show how FAD can bring an answer to some of them. For instance one can wonder whether the local dynamics in the melt are affected by chain connectivity at they are in dilute solution, or if the numerous interchain constraints lead to a completely different behavior. Moreover, the relation between molecular motions and macroscopic properties is rather unknown.

## 3 Recent Progress in the Fluorescence Anisotropy Decay Technique

### 3.1 Principle of Fluorescence Anisotropy Decay

At usual temperatures the electrons of a molecule occupy the orbitals of lower energy. Under the action of an electromagnetic field (generally light in the U.V. range) one electron may jump to an orbital of higher energy by absorption of a photon. The molecule is then excited. The exicited state has a finite lifetime (generally a few nanoseconds), and the molecule may return to the fundamental state by emission of a fluoresence photon at a wavelength greater than the absorption wavelength. The decay of fluorescnece intensity is roughly exponential.

The absorption of light is proportional to the scalar product of the incident electric field and of a molecular vector named the transition moment. Thus, excitation of an isotropic population of fluorescent species by polarized light generally creates a temporary anisotropic population of excited molecules. Molecular motions progressively destroy this anisotropy, and affect the polarization of the reemitted fluorescence light which can be studied using pulse fluorometry techniques.

The interesting quantity, the fluorescence anisotropy, is defined by:

$$r = \frac{I_V - I_H}{I_V + 2 I_H}$$

where $I_V$ and $I_H$ correspond to fluorescence intensities for analyzer direction parallel and perpendicular respectively to the vertical polarization of the incident beam. In this expression $(I_V + 2 I_H)$ represents the total fluorescence intensity.
— The anisotropy decay following an infinitely short pulse at time zero,

$$r(t) = [I_V(t) - I_H(t)]/[I_V(t) + 2 I_H(t)] \tag{10}$$

is proportional to the 2$^{nd}$ moment of the OACF of the emission transition moment:

$$r(t) = r_0 M_2(t)/M_2(O) \tag{11}$$

The fundamental anisotropy $r_0$ is a time-independent molecular parameter. Thus, the OACF can be sampled quasi-continuously directly in the time domain, provided one can sample precisely r(t). This properties makes FAD a rather unique tool for discussing the different models for the OACF. (In principle, the transient Kerr Effect is also able to provide such a sampling, but, at the present time, this latter techniques does not seem to reach the same precision).

## 3.2 The FAD Apparatus

FAD is now a well-known technique, and the general features of an experiment have been explained in full details by several authors [49,50]. We briefly present here the apparatus used in our studies, which has already been described elsewhere [51]. Its block diagram is given in Fig. 1. The excitation wavelength is selected from synchrotron radiation by a double holographic grating monochromator. It is vertically polarized. The emission wavelength is selected by a single holographic grating monochromator.

We use single photon counting [49,52], i.e. no more than one fluorescence photon is received by the photomultiplier for one excitation pulse. On receiving this photon, the cathode of the PM ejects a photoelectron which is amplified in order to produce an electric pulse at the anode. This pulse is sent to the "start" inject of a time to amplitude converter (TAC) at a time $t_1$, which is compared with the time $t_2$ at which the belayed reference pulse is received by the "stop" input. At its output, the TAC sends a pulse whose amplitude is proportional to $t_2 - t_1$ to an analogic-digital

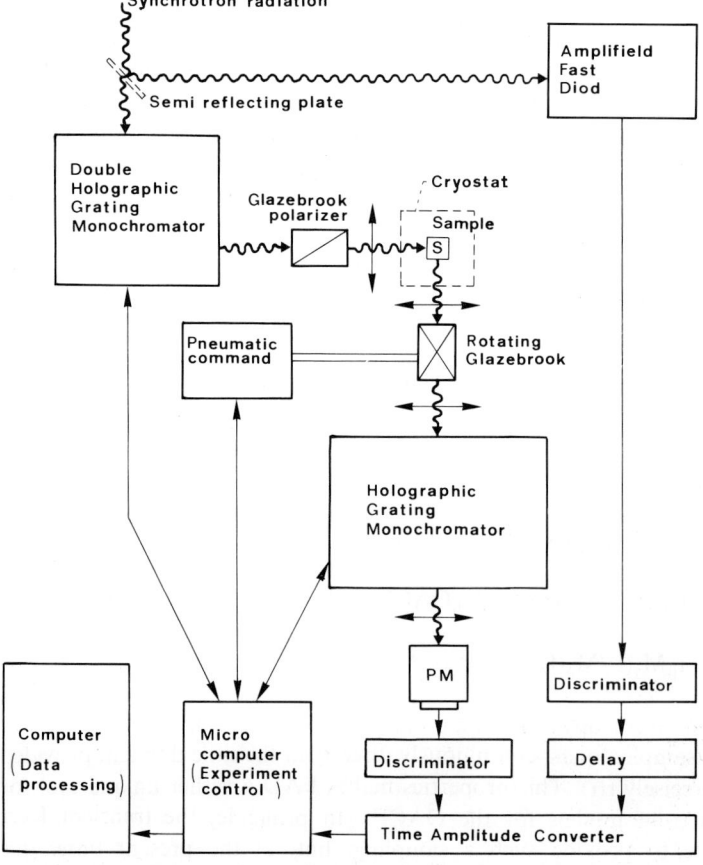

**Fig. 1.** Block diagram of the Fluorescence Anisotropy Decay experiment

couverter and then to the computer, which stores one count in the channel corresponding to the time $t_0 = t_2 - t_1$.

After many excitation cycles, the number of counts which have been accumulated in one channel is proportional to the probability of emission at a given time after excitation. The errors on these number of counts are random, independent and follow the well known Poisson distribution [33]. The emission decay is recorded alternatively in vertical (Parallel) and horizontal (Perpendicular) polarizations using a computer-commanded rotating Polarizer.

At last, the finite width of the excitation pulse is accounted by a non linear least square iterative reconvolution procedure. Thus, a good precision on r(t) can be achieved if:

(i) The counting rate is high, to minimize the random relative error proportional to: $n_i^{-1/2}$, where $n_i$ is the number of counts in channel i;

(ii) One remains in the single photoelectron regime;

(iii) The excitation pulse is as short as possible, as stable in time as possible, and correctly sampled. Indeed, synchrotron radiation fulfills these requirements much better than conventional light sources.

## 3.3 The Synchrotron Radiation

Initially built up for the study of elctron-positron collisions, cyclosynchrotrons rapidly appeared as unique sources for the production of electromagnetic radiation in the X-ray and UV range [54-55]. At the present time, several machines are dedicated to the production of Synchrotron radiation, as the one we have used, LURE-ACO at Orsay (France) [55] (Fig. 2).

In a synchrotron, an electron (or positron) bunch issued from a linear accelerator is injected into a vacuum ring chamber and maintained colinear to the axis of the

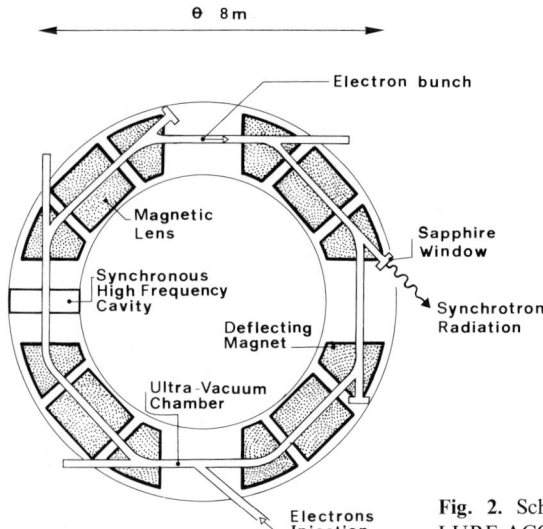

Fig. 2. Schematic view of the cyclosynchrotron LURE-ACO (Orsay, France)

chamber by intense magnetic fields. In the curved parts of the ring, the relativistic electrons emit pulsed light with a continuous spectrum ranging from X ray to infra-red. The width and shape of the light pulses depends on the size of the electron bunch but not on the wavelength. It can range from 50 picoseconds to several nanoseconds depending on the characteristics of the synchrotron, and it is very stable in time. In ACO, the pulse width is about 1 ns, and the intensity decays exponentially with a typical lifetime of 10 hours or more. These characteristics can be compared to those of an usual flash lamp in Table 2.

The light flux is at least two orders of magnitude greater than that of conventional pulsed flash sources and the very high repetition rate $v_e \simeq 13$ MHz enables the use of high fluorescence counting rates $v_f$ in the single photon regime ($v_f < 10^{-2} v_e$). Indeed, even with sharp monochromation ($\Delta\lambda = 4$ nm) both in excitation and emission, $v_f$ is mainly limited by the response of the electronics. For decays longer than the period of ACO (73 ns), the decays following successive pulses partly overlap. This overlap can be accounted for in the reconvolution procedure, and a satisfying determination of correlation times up to about 200 ns can be achieved.

The continuous spectrum of the synchrotron allows a direct sampling of the excitation pulse at the emission wavelength, avoiding the perturbations due to the

**Table 2.** Comparison between the performances of LURE-ACO and those of a classical flash lamp

| Characteristics | Flash lamp | Synchrotron Radiation ACO-LURE (Orsay — France) |
|---|---|---|
| Pulse width | $\simeq 2$ ns | $\sim 1$ ns |
| Pulse shape | Asymmetric | gaussian |
| Pulse shape stability | <0.1 ns/h | Long time evolution (0.02 ns/h) |
| Intensity variation | Random fluctuation (several percent) | Exponential decay (Cte $\sim$ 10 h) |
| Max. counting rate in single photon regime | 400 c · s$^{-1}$ | 15 000–500 000 c · s$^{-1}$ |
| Required counting time | 1–10 h | 10 min. |
| Time interval between pulses | >10 µS. | 75 ns |
| Excitation wavelengths | Discrete set or continuous spectrum | Continuous spectrum (X ray — IR) |
| Correction of P. M. characteristics with wavelength | Approximate | Direct calibration from scattering at emission wavelength |
| Intensity calibration for $I_V$, $I_H$ | Accuracy $\approx 10^{-2}$ | Accuracy $\approx 10^{-3}$ |
| Absolute accuracy on r | $|\Delta r| = 0.01$ | $|\Delta r| = 0.01$ with emission monochromator $|\Delta r| = 0.004$ with emission optical filter |
| Correlation times available | | |
| minimum | $\approx 0.5$ ns | $\approx 0.1$ ns |
| Maximum | $\approx 50$–400 ns depending on fluorescence lifetime | $\approx 200$ ns |
| Accessibility | Easy | Difficult (yearly schedule, proposals) |

wavelength-dependence of the P. M. response [56-61], which affects the precision of the deconvolution.

All these characteristics allow to measure FAD with a statistic quality and a reliability out of reach of flash sources. Moreover, in the case of polymers, it is generally not possible to purify samples as much as one would wish to perform a fluorescence experiment in comfortable conditions, and the free choice of wavelength permitted by the continuous spectrum of the synchrotron source is essential. In this regard, lasers, which also provide very intense and short light pulses usable for fluorescence experiments [62-67], are somewhat less flexible. This may partly explain why, to our knowledge, laser systems have not been applied to polymer dynamics until now.

## 3.4 Quantitative Criteria for Data Interpretation

As quoted in Sect. 1.2, different theoretical models for chain dynamics generally lead to different analytical expression for the OACF, which can be compared with the experimental anisotropy. The knowledge of the statistical distribution of errors on each channel is an essential tool for this comparison. It provides objective criteria to decide whether a discrepancy between a model and a set of data is significant or not, and to compare different models. Among these criteria, the most well known one is the reduced $\chi^2$, [53], which should be 1 for purely statistical deviations, and increases with increasing systematic discrepancy. However, several other statistical criteria can be used [53,57,68]. We recently developed [37] a systematic method for discussing theoretical expressions, using statistical criteria, together with physical criteria aimed to discriminate expressions which may fit the data correctly but lead to parameters which behave in a non-physical way. For instance, if a model is supposed to reflect the true OACF of the chain backbone, the best fit parameters should not vary significantly when this OACF is sampled with a different experimental window.

Also, the fundamental anisotropy $r_0$ cannot exceed 0.4, for theoretical reasons.

Of course, the comparison with experiments by a single technique cannot prove that a model is the unique realistic representation of molecular motions. But, in Sec. 4, we give some examples in which the aforementioned procedure has been useful to select between the different expressions proposed in literature and improve our knowledge of chain dynamics.

## 3.5 Fluorescent Labeling of Polymer Chains

In order to study the dynamics of a polymer chain, one must covalently bound a fluorescent label in the middle of the chain.

In the method proposed by Valeur [38], monofunctional "living" monodisperse chains are prepared and deactivated by 9–10-bis(bromomethyl)anthracene. The resultant chains contain dimethyl anthracene in their middle, as shown in Fig. 3 (This figure represents labelled polystyrene, but other polymers can be labelled too). Anthracene is a particularly convenient label since it is rigid and rather small, it has a good quantum yield and it is easy to excite. When it is bounded in 1,9 positions, its

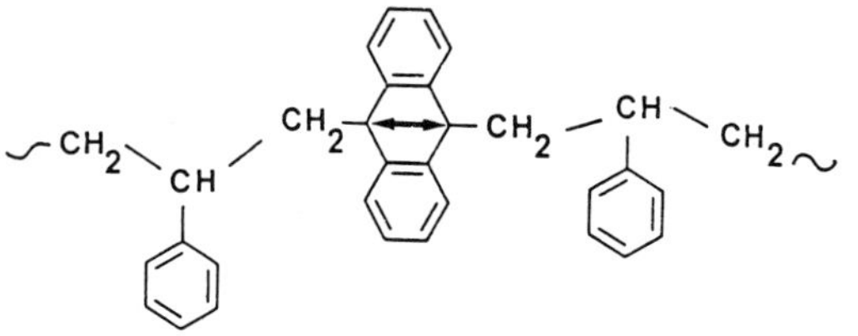

**Fig. 3.** Polystyrene labelled with anthracene in the middle of the chain (PSAPS)

transition moment is parallel to the chain axis, and its internal rotation about the 1,9 axis has no action on the OACF of its transition moment. We have always found that the mean correlation time of the DMA label coupled in the middle of a polymer chain in a given matrix was more than one order of magnitude larger than the correlation time of free DMA in the same matrix. This observation supports the assumption that the motion of the transition moment in the labelled chain is mainly controlled by chain dynamics.

Other information can be obtained from fluorescent probes physically inserted in the polymer matrix [19,20]. This method is not discusses here, since it does not lead to a direct access to polymer-segment correlation functions.

# 4 Use of Fluorescence Anisotropy Decay for Studying the OACF of Polymers

## 4.1 Dilute Solutions of Labelled Chains

Since the OACF of chain bonds in bulk polymers has practically never been studied from an experimental of theoretical point of view, we briefly recall the most recent results obtained for the OACF in dilute solutions to guide us in the approach of bulk polymers. At the same time, we check how our improved FAD method can be compared with other techniques in that case, and how it is able to tell apart the different theoretical models. For the sake of simplicity, we present here the results obtained for a well known polymer: Polystyrene labelled with anthracene in the middle of the chain [13] PSAPS, Fig. 3. This polymer was studied at 25 °C in dilute solution in different mixtures of ethylacetate and tripropionin (glyceryl tripropionate) in order to vary the viscosity in a wide range of values (from 0.4 to 8 cp). For each solution, the different models introduced in Sect. 2.2 and 2.3 were fitted to the data. This comparison once again shows the definitely non exponential character of the OACF. The monoexponential best fit to one of the experimental anisotropies (recorded at a viscosity $\eta = 5.4$ cp), presented in Fig. 4, shows a systematic deviation clearly apparent in the weighted residuals (upper curve). The reduced $\chi^2$ are also much larger than 1, as can be checked on the examples given in Table 3. On the contrary, the 3 expressions WW, BY and HH lead in any case to a fit which is satisfying from a curve-fitting point of view (see Fig. 5 and 6). In these cases the weighted residuals are randomly distributed. Only the VJGM model always leads to high $\chi^2$ and poor fit.

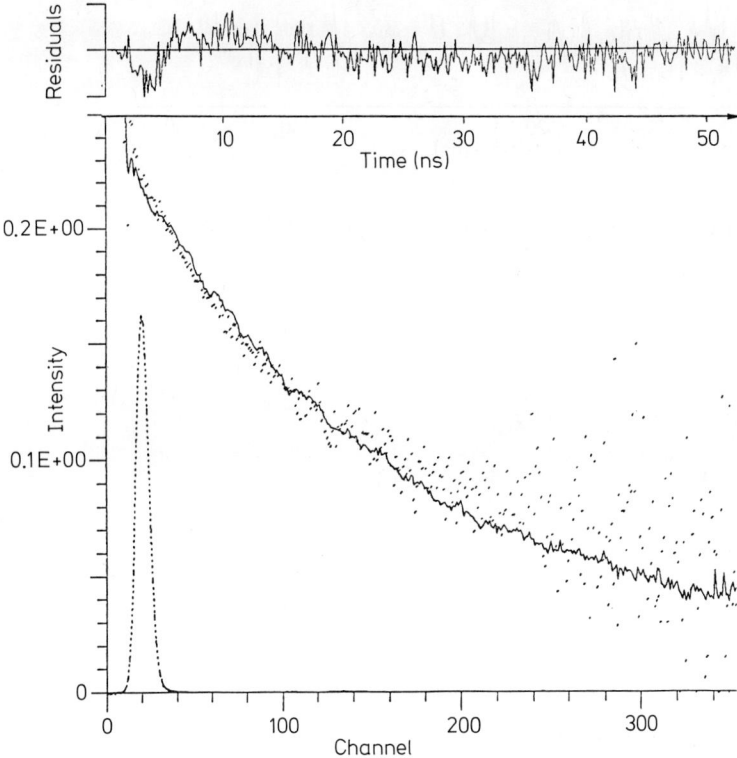

**Fig. 4.** Comparison of the isotropic rotation model to the experimental anisotropy of PSAPS in solution ($\eta = 5.4$ cp). Dots are experimental data. The continuous line is the best fit OACF reconvoluted by the measured instrumental function (exciting pulse). This pulse is plotted as a dash-dot line (arbitrarily scaled). The upper graph represents the weighted residuals

**Table 3.** Reduced $\chi^2$ for different samples and different models

| Model | Viscosity cp | | |
|---|---|---|---|
| | 0.43 | 2.28 | 5.40 |
| IR | 2.56 | 3.22 | 1.83 |
| RR | 1.33 | 1.21 | 1.13 |
| WW | 0.97 | 1.09 | 0.98 |
| VJGM | 1.44 | 1.25 | 1.05 |
| BY | 1.06 | 1.11 | 1.00 |
| HH | 1.06 | 1.14 | 1.02 |

This result is not surprising, since this model predicts an unrealistic infinite slope at time zero. Nevertheless, to our knowledge, it is the first time that this theoretical weakness is demonstrated by experimental results.

As quoted in Sect. 3.4, this discussion of models can be improved by physical criteria. For instance, we have shown that the satisfying $\chi^2$ values obtained with the

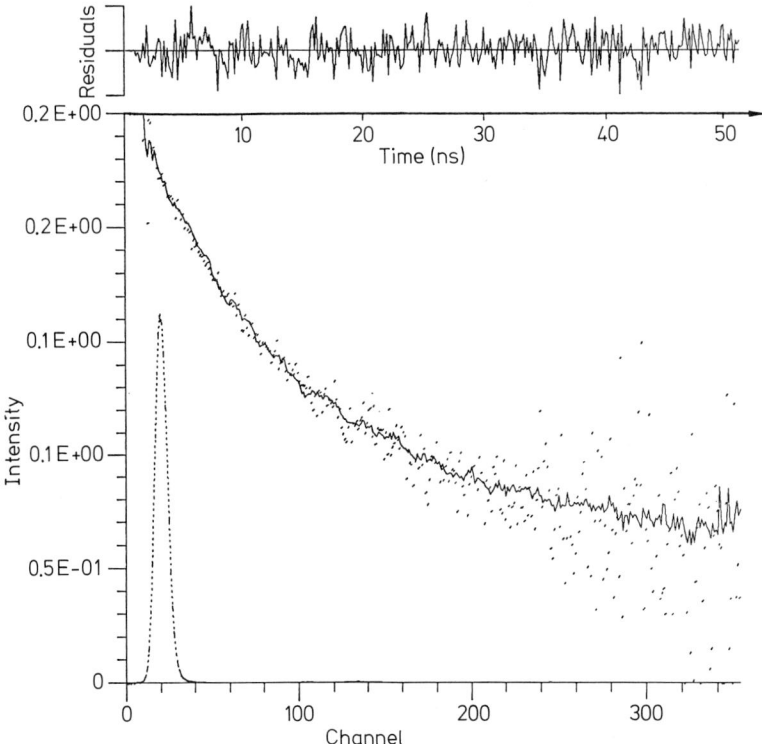

**Fig. 5.** Comparison of the WW expression to the experimental anisotropy of PSAPS (same representation as Fig. 4)

WW model correspond to parameters which are not physically reasonable and stable as regards to variations in the experimental window (see Table 4).

Similar conclusions can be drawn when one uses models describing the anisotropic motion of rigid bodies: good curve fitting may be obtained fortuitously, but, in contrast with the case of the BY on HH expressions, this fitting does not lead to stable and significant parameters.

These results were corroborated by studies of labelled polystyrene in other solvents as toluene, styrene and pure tripropionin, at different temperatures. The main conclusions of these studies of polymer dynamics in dilute solution are the following ones:

— the connectivity and flexibility of the chains lead to theoretical OACF which present unique features. Indeed, these features are observed experimentally. Only specific models taking into account the molecular nature of polymer chains are able to account for the experimental OACF.
— among the different closed expressions derived from specific polymer models, further distinctions can be made, and only the BY and HH models remain acceptable at the present level of experimental precision. But, as quoted in Sect. 2.2, the arbitrary cutoff frequencies in the B.Y model are difficult to interpret on a molecular scale.
— one can notice that the HH expression is simply the product of a diffusive term with characteristic time $\tau_1$ and of a loss term with characteristic time $\tau_2$. Since

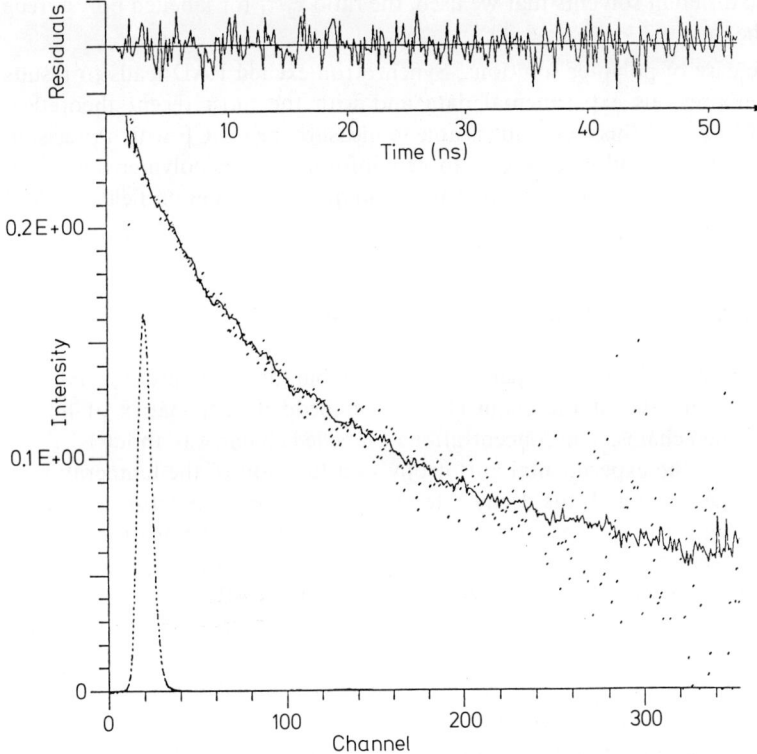

**Fig. 6.** Comparison of the HH expression to the experimental anisotropy of PSAPS (same representation as Fig. 4)

**Table 4.** One example of the best-fit parameters obtained when different models are fitted to the experimental anisotropy of PSAPS (viscosity 0.43 cp)

| Model | $r_0$ | $\tau_1$ ns | $\tau_2$ ns | $\tau_2/\tau_1$ (or $\beta$) | $\chi^2$ |
|---|---|---|---|---|---|
| RR   | 0.166 | 2.76  |      | $\beta = 0.018$ | 1.337 |
| WW   | 0.725 | 1.81  |      | $\beta = 0.712$ | 0.972 |
| VJGM | 0.171 | 147   | 3.26 | 0.002 | 1.436 |
| BY   | 0.200 | 0.857 | 8.40 | 9.8   | 1.057 |
| HH   | 0.199 | 2.55  | 6.71 | 2.62  | 1.062 |

these 2 processes seem to be a very general feature of polymer dynamic models, the agreement between the HH model and experiment is a satisfying observation. Moreover, it suggests that this expression can be used rather extensively to analyse experimental data, even if the molecular origin of the two processes is still to be precised.

- FAD is able to measure precisely the characteristic time of the diffusive process, $\tau_1$, and, in some cases, the characteristic time $\tau_2$ of the loss process (with a lower precision). The ratio $\tau_2/\tau_1$ seems to depend on the solvent, but not on the temperature. In the different solvents that we used, the ratio $\tau_2/\tau_1$ for labelled polystyrene varies in the range: $3 < \tau_2/\tau_1 < 30$.

Thus, in the case of polymer solutions, Synchrotron excited FAD leads to results consistent with previous experimental data and with the most recent theoretical expectations. Moreover, this technique is able to measure the OACF with a precision unavailable previously, and gives access to new information on polymer dynamics. These observations support the use of this technique in the newer field of local dynamics in bulk polymers.

## 4.2 OACF of Labelled Chains in a Bulk Polymer

We recently performed [37] FAD experiments on polybutadiene chains labelled with anthracene in the middle of the chain (Fig. 7), embedded in a matrix of homo-polymers unlabelled chains. The concentration of labelled chains was about 1%.

The evolution of the experimental anisotropy as a function of the temperature is shown in Fig. 8. As expected, the decay rate increases as the temperature increases. For the highest temperature (t > 50 °C), it can be noticed that the anisotropy decays from a value close to the fundamental anisotropy of DMA to almost zero in the time window of the experiment (about 60 ns). This means that the initial orientation of a backbone segment is almost completely lost within this time. This possibility to directly check the amplitude of motions associated with the involved relaxation is a very useful advantage of FAD. In particular, it indicates that in the temperature range 50 °C ~ 80 °C, we sample continuously and almost completely the elementary brownian motion in polymer melts. Processes too fast to be observed by this technique involve only very small angles of rotation and cannot be associated with backbone rearrangements. On the other hand, the processes too slow to be sampled concern only a very low residual orientational correlation, i.e. they are important only on a scale much larger than the size of conformational jumps.

The different available expressions for the OACF (Table 1) were compared to the data. As for solutions, the non-polymer models (rotational diffusion or restricted rotation (RR)) gives a poor fit. This is shown in Fig. 9 where the best fit OACF for the RR model is compared with the experimental data ($\Theta = 62.7$ °C). Thus, the very specific character of main-chain orientation relaxation is as apparent in melt polymers as it was in solution.

But we have also observed that the WW expression leads to very unstable best fit parameters, and non-realistic values for $r_0$. Even from a mere curve-fitting point of view, this expression is not as good as the others (compare Figs. 10 and 11). As an

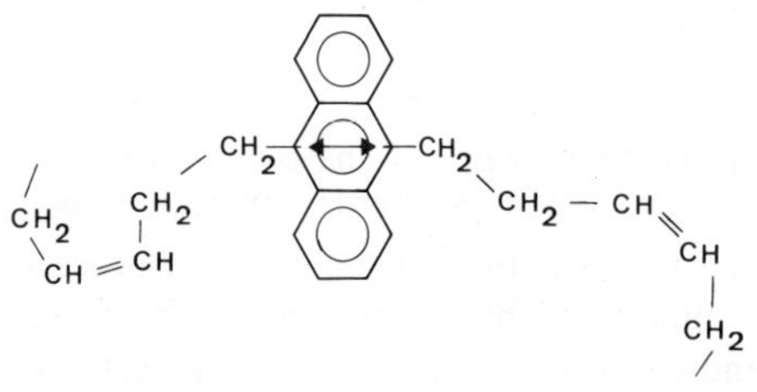

Fig. 7. Polybutadiene labelled with anthracene in the middle of the chain (PBAPB)

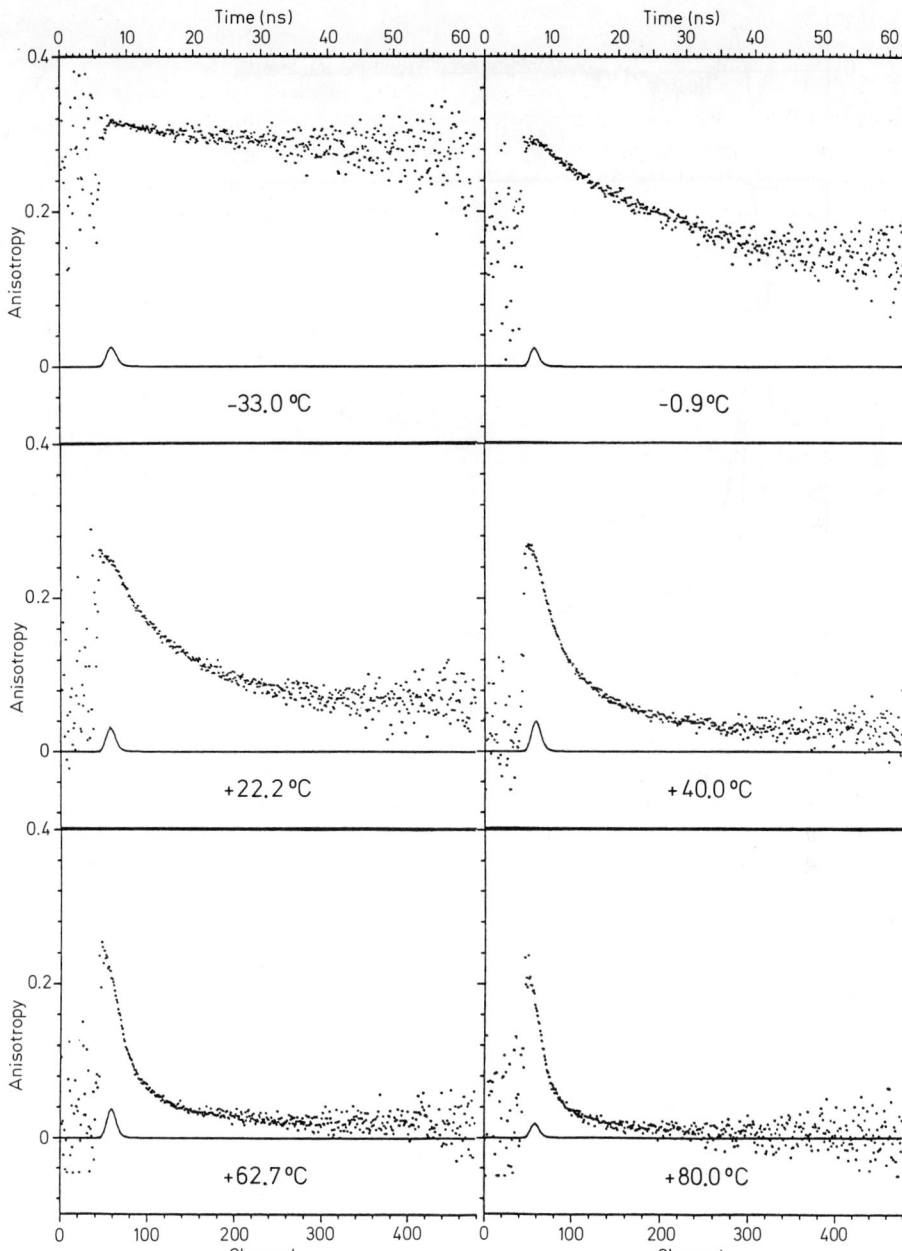

**Fig. 8.** Evolution of the FAD of PBAPB in bulk PB as a function of temperature

example the best fit parameters obtained for the different models at 62.7 °C are given in Table 5.

Thus, the WW expression, which appeared to account for the distribution of relaxation times observed in multimolecular experiments [69-72], does not correspond

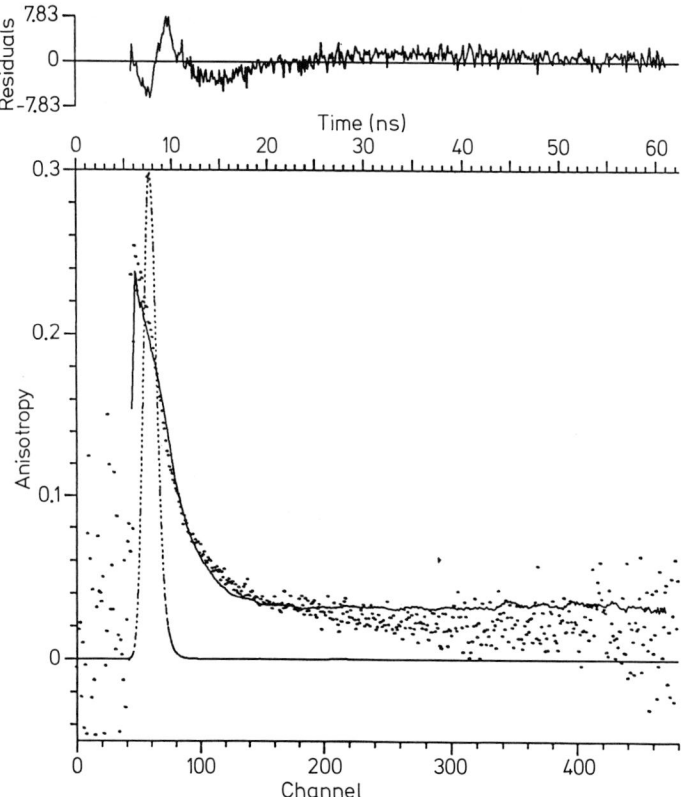

**Fig. 9.** Comparison of the RR model to the experimental anisotropy of PBAPB at 62.7 °C (same representation as Fig. 4)

to the shape of the OACF for one given vector of the chain. This observation may be surprising at first sight. But, as emphasized in Sect. 2.1, multimolecular experiments involve the time evolution of many intrachain and interchain correlations. Thus, one should expect significant differences from the OACF probed in FAD.

On the other hand, the recent models which accounted well for the dynamic behavior in solution (HH model and related [35-37]) seem to account also for the OACF in melt polymers (see Table 5). This rather good fit can also be checked by direct visual observation of Fig. 11 (The small discrepancy observed in channel 20 to 40 corresponds to the exciting pulse. It may betray a residual diffused stray light, and the difficulty to perform precise optical experiments in bulk polymers).

Moreover, the ratios $\tau_2/\tau_1$ obtained at different temperatures are similar to the values obtained for labelled polybutadiene in dilute solution in toluene ($\simeq 30$). Thus, the surrounding chains seem to affect local dynamics only at the level of the friction coefficient. Intrachain connectivity remains the essential non isotropic constraint on local dynamics in bulk polymers and models for the isolated chain are applicable. This is consistent with the idea that topological interchain effects only act on a

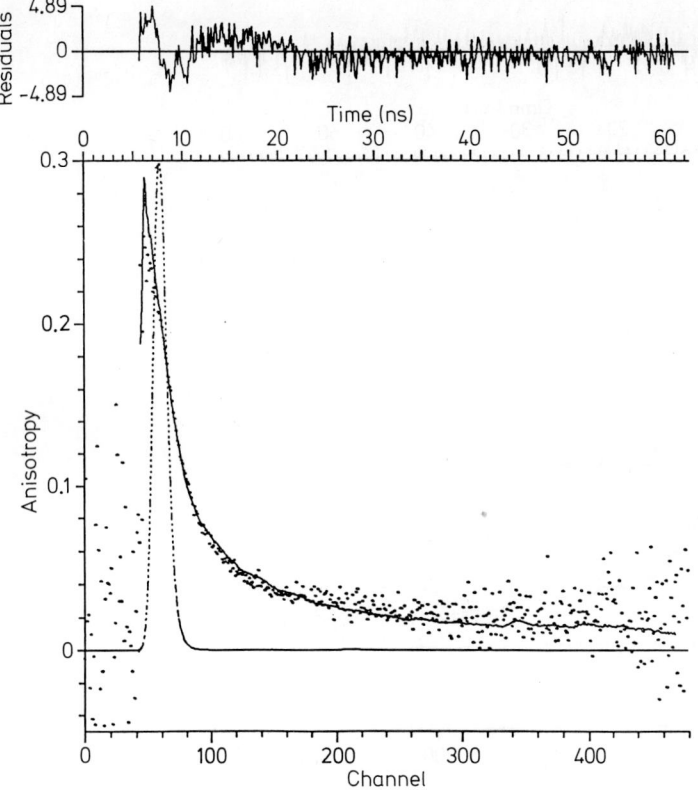

**Fig. 10.** Comparison of the WW expression to the experimental anisotropy of PBAPB at 62.7 °C (same representation as Fig. 4)

large scale [73–75] (entanglement distance) and do not change to local dynamic processes.

From a practical point of view, these observations are important for techniques such as ESR or NMR, which require the a priori choice of a model to be interpreted.

## 4.3 Study of temperature Effects

The macroscpic dynamic properties of polymers have a well known but rather poorly understood temperature dependence, and it is very important to study the corresponding temperature evolution of molecular dynamics.

The HH model fits correctly experimental anisotropies in all the temperature range explored (−55 °C/80 °C). But the long time loss term rapidly falls far off the experimental window when the temperature is decreased, and its characteristic time $\tau_2$ could be determined with a reasonable precision only above 40 °C. In the range 40 °C/80 °C, the ratio $\tau_2/\tau_1$ does not seem to vary significantly, and remains rather high ($\simeq 30$). If one accepts the separation of the OACF into a one-dimension

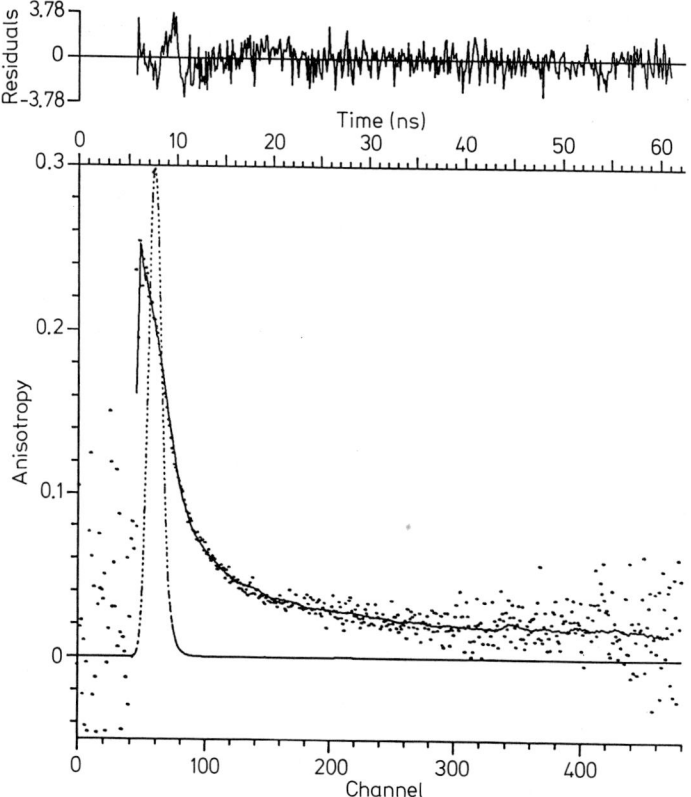

**Fig. 11.** Comparison of the HH expression to the experimental anisotropy of PBAPB at 62.7 °C (same representations as Fig. 4).

diffusion term and a loss term, this observation implies that the processes responsible for the loss (whatever their molecular nature is) are slow as regards to be diffusive ones. Indeed, when the HH expression is used without loss ($\tau_2 = \infty$), fitting remains rather satisfying and $\chi^2$ increases only slightly.

The values of $\tau_1$, plotted on Fig. 12 leads to an apparent activation energy of 40 ± 5 KJ/mole. To compare these results with macroscopic relaxation, we used the well-known WLF [2,6] time-temperature superposition equation. According to this equation, the value of the principal or "glass transition" relaxation time $\tau_a$ at a temperature $T_a$ can be deduced from its value $\tau_b$ at $T_b$ following:

$$\log\left(\frac{\tau_a}{\tau_b}\right) = \frac{C_1^0(T_b - T_g)}{C_2^0 + T_b - T_g} - \frac{C_1^0(T_a - T_g)}{C_2^0 + T_a - T_g} \qquad (12)$$

Where $T_g$ is the glass transition temperature and $C_1^0$ and $C_2^0$ are phenomenological parameters. In figure 8, we have arbitrarily chosen for $\tau_b$ the value of $\tau_1$ at 40.9 °C, i.e. 4.54 ns, and applied Equ. (12) together with the parameters $C_1^0$ and $C_2^0$ given by Ferry [2] from low frequency mechanical measurements. The corresponding "WLF curve" (dotted line) fits well the other measured values of $\tau_1$. Of course, a similar

**Table 5.** Best fit parameters obtained when different models are fitted to the anisotropy of PBAPB at 62.7 °C using different experimental windows 1:0 ns — 55 ns; 2:1.3 ns — 55 ns; 3:3.2 ns — 55 ns; 4:0 ns — 37 ns; 5:0 ns — 17 ns

| Model | Truncation | $\chi^2$ | $r_0$ | $\tau_1$ ns | $\tau_2$ ns (or $\beta$) | $\tau_2/\tau_1$ |
|---|---|---|---|---|---|---|
| VJGM | 1 | 1.9709 | 0.336 | 1.07 | 30.6 | 28.6 |
|  | 2 | 1.774 | 0.345 | 0.99 | 31.5 | 31.8 |
|  | 3 | 1.134 | 0.424 | 0.60 | 29.1 | 48.5 |
|  | 4 | 2.416 | 0.322 | 1.13 | 28.9 | 25.6 |
|  | 5 | 2.2754 | 0.311 | 1.61 | 19.5 | 11.7 |
| WW | 1 | 1.8650 | 0.333 | 1.568 | 0.394 |  |
|  | 2 | 1.7213 | 0.347 | 1.40 | 0.381 |  |
|  | 3 | 1.299 | 0.596 | 0.28 | 0.275 |  |
|  | 4 | 2.200 | 0.330 | 1.61 | 0.400 |  |
|  | 5 | 1.0017 | 0.307 | 1.98 | 0.446 |  |
| HH | 1 | 1.2477 | 0.234 | 1.222 | 96.8 | 43.6 |
|  | 2 | 1.233 | 0.234 | 2.22 | 101.1 | 45.5 |
|  | 3 | 1.1053 | 0.212 | 2.73 | 89.0 | 32.6 |
|  | 4 | 1.434 | 0.234 | 2.23 | 95.2 | 42.6 |
|  | 5 | 1.8238 | 0.234 | 2.26 | 83.8 | 37.1 |
| BY | 1 | 1.1917 | 0.242 | 0.641 | 151 | 235 |
|  | 2 | 1.1773 | 0.242 | 0.638 | 152 | 238 |
|  | 3 | 1.168 | 0.230 | 0.718 | 139 | 193 |
|  | 4 | 1.327 | 0.242 | 0.647 | 144 | 222 |
|  | 5 | 1.4664 | 0.240 | 0.681 | 112 | 164 |

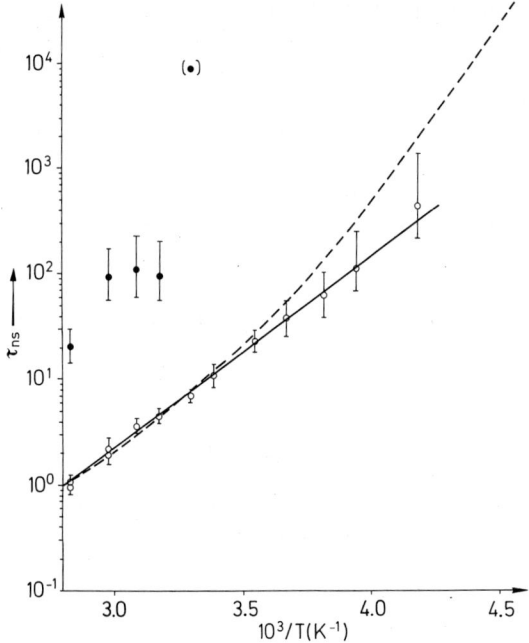

**Fig. 12.** Arrhénius plot of the best fit parameters $\tau_1$ (○) and $\tau_2$ (●) (HH model) for PBAPB unbedded in a PB matrix. The dashed line is the WLF curve according to Ref. [2]

procedure can be applied to $\tau_2$, but the poorer precision on this parameter makes it less significant.

The local reorientation processes observed in FAD and the macroscopic relaxation have the same temperature behavior. Similar observations have already been made using NMR or Fluorescence polarization under continuous excitation [19,20]. However, for these techniques, the slopes of the curves depend on the choice of the model [20], so that the confidence one could put in the agreement (or discrepancy) between spectroscopic and mechanical results relies directly on the confidence one put in the model arbitrarily chosen.

In the present set of experiments, we have first demonstrated that the model used to interpret the data accounted precisely for the orientation relaxation in a rather wide range of temperatures. Thus, our results strongly support the idea that the local reorientation processes observed in FAD are indeed the elementary processes of macroscopic viscoelasticity.

# 5 Conclusions

FAD allows to study rather precisely and unambiguously the OACF of polymer chains. Thanks to the unique statistical nature of the single photon method, and to the performances of the synchrotron source, it has proved very useful in the discussion of the nature of motions. For instance, it led to the first observation of the diffusion of orientational motions along the chains in polymer melts. Of course, the labeling nature of this kind of experiments implies two limitations. The first one, technical, is the necessity of labeling chains. Indeed, many different polymers can be labelled and several species have been or are studied in our laboratory (Polystyrene [13,32,38], polybutadienes [19,20,37], polyisoprenes [19,20], polyethylene [76]. But labeling limits the use of FAD as a routine technique.

The second limitation, fundamental, is the local perturbation of the chain induced by the label. From that point of view, comparison with the NMR technique is very interesting. NMR probes the motions of well-defined vectors in the chain, without any perturbation. In contrast to FAD, it is able to give quantitative information about subtle effects such as the influence of local conformations on dynamics. But this information is contained in a few integral quantities ($T_1$, $T_2$, NOE, at 2 or 3 frequencies in the best cases) which are also affected by molecular motions on a larger scale. The error on these quantities is not given by a simple statistic law as in FAD, so that NMR alone often leads to the unsatisfying situation where very different molecular interpretations can be given to the same experiment.

From these somewhat oversimplified considerations, it appears that techniques like NMR and FAD are not concurrent but complementary. They explore motions in a different way and on overlapping but different distance ranges. We agree with the fact that non perturbing technique like NMR contain the most precise information about local molecular dynamics. But we also believe that the knowledge of motions on a scale intermediate between the atomic bond and the entanglement length, which is given by Synchrotron excited FAD, will be essential to obtain this information unambiguously.

# 6 References

1. Alfrey, Jr. T.: Mechanical Behavior of High Polymer, Interscience, New York, 1948
2. Ferry, J. D.: "Viscoelastic properties of Polymers" (2 nd ed.), Wiley, New York, 1970
3. Ishida, Y.: J. Polym. Sci., A 2, 7, 1835, (1969)
4. "Dielectric properties of Polymers", F. E. Karasz éd., Plenum Press, London, 1972
5. Williams, M. L.: J. Phys. Chem., 59, 95 (1955)
6. Williams, M. L., Landel, R. F., Ferry, J. D.: J. Amer. Chem. Soc., 77, 3702 (1955)
7. Cohen, M. H., Turnbull, D.: J. Chem. Phys., 31, 1164 (1959), 34, 120 (1961); 52, 3038 (1970)
8. Gibbs, J. H., di Marzio, E. A.: J. Chem. Phys., 28, 373 (1958); 28, 807 (1958)
9. Adams, G., Gibbs, J. H.: J. Chem. Phys., 43, 139 (1965)
10. Cohen, M. H., Grest, G. S.: Phys. Rev. B., 20, 1077 (1979)
11. Gibbs, J. H., di Marzio, E. A.: J. Polym. Sci., 40, 121 (1959); A1, 1417 (1963)
12. Friedrich, C., Laupretre, F., Noël, C., Monneric, L.: Macromolecules 13, 1625 (1980); 14, 119 (1981)
13. Viovy, J. L., Monnerie, L., Brochon, J. C.: Macromolecules, 16, 1845 (1983)
14. Spiess, H. W.: J. Chem. Phys. 72, 6755 (1980)
15. Lindler, P., Rossler, E., Sillescu, H.: Makromol. Chem., 182, 8653 (1981)
16. Howarth, O. W.: J. Chem. Soc. Faraday Trans. II, 76, 1219 (1980)
17. Gronski, W., Murayama, N.: Makromol. Chem., 180, 277 (1979)
18. Törmala, P.: J. Macromol. Sci., Rev. Macr. Chem., C17, 197 (1979)
19. Jarry, J. P., Monnerie, L.: J. Polym. Sci., Polym. Phys. Ed., 16, 443 (1978); 18, 1879 (1980)
20. Queslel, J. P.: Thesis, Paris, 1982
21. Bower, D. I.: J. Polym. Sci., Polym. Phys. ed., 19, 93 (1981)
22. Debye, P.: "Polar Molecules", Chemical Catalog Co., New York (1929)
23. Perrin, F.: Ann. Phys., 12, 169 (1929)
24. Perrin, F.: J. Phys. Le Radium, 5, 497 (1934); 7, 1 (1936)
25. Favro, L. D.: Phys. Rev., 119, 53 (1960)
26. Hu, C. M., Zwanzig, R.: J. Chem. Phys., 60, 4354, (1974)
27. Youngren, G. K., Acrivos, A.: J. Chem. Phys., 63, 3846 (1975)
28. Kinosita, K., Kawato, S., Ikegami, A.: Biophysical Journal, 20, 289 (1977)
29. Wahl, P.: Chem. Phys., 7, 210 (1975)
30. Rouse, P. E.: J. Chem. Phys., 21, 1272 (1953)
31. de Gennes, P. G.: Physics (Long Island City N. Y.), 3, 37 (1967)
32. Valeur, B., Jarry, J. P., Geny, F. Monnerie, L.: J. Polym. Sci., Polym. Phys. Ed., 13, 667 (1975); 13, 675 (1975)
33. Bendler, J. T., Yaris, R.: Macromolecules, 11, 650 (1978)
34. Jones, A. A., Stockmayer, W. H.: J. Polym. Sci., Polym. Phys. Ed., 15, 847 (1975)
35. Hall, C. K., Helfand, E.: J. Chem. Phys., 77, 3275 (1982)
36. Weber, T. A., Helfand, E.: J. Phys. Chem. 87, 2881 (1983)
37. Viovy, J. L., Monnerie, L., Merola, F.: Submitted to Macromolecules
38. Valeur, B., Monnerie, L.: J. Polym. Sci., Polym. Phys. Ed., 14, 19 (1976); 14, 29 (1976)
39. Laupretre, F., Noël, C., Monnerie, L.: J. Polym. Sci. Polym. Phys. Ed., 15, 2127 (1977)
40. Heatley, F., Wood, B.: Polymer, 19, 1405 (1978)
41. Gronski, W., Schafer, T., Peter, R.: Polym. Bull. 1, 319 (1979)
42. Tekely, P., Turska, E.: Polymer, 24, 667 (1983)
43. Allen, G., Brier, P. N., Goodyear, G., Higgins, J. S.: Faraday Symp. Chem. Soc., 6, 169 (1972)
44. Allen, G., Higgins, J. S., Wright, C. J.: J. Chem. Soc. Faraday Trans II, 70, 348 (1974)
45. Allen, G., Gosh, R. E., Heidemann, A., Higgins, J. S., Howells, W. S.: Chem. Phys. Lett., 27, 308 (1974)
46. Higgins, J. S., Gosh, R. E., Howells, W. S., Allen, G.: J. Chem. Soc. Faraday Trans II, 73, 40 (1977)
47. Higgins, J. S., Nicholson, L. K., Hayter, J. B.: Polymer, 22, 163 (1981)
48. Williams, G., Watts, D. C.: Trans Faraday Soc., 66, 80 (1971)
49. Wahl, P.: New Tech. Biophys. Cell. Biol. 2, 233 (1975)
50. Wahl, P.: "Decay of Fluorescence Anisotropy" in "Concepts in Biochemical Fluorescence", (R. F. Chen and H. Edelhoch Eds.), Marcel Dekker, New York, 1975

51. Brochon, J. C.: in "Protein Dynamics and Energy Transduction" Shin'ichi Ishiwata, Ed. Taniguchi Foundation, Japan, 1980
52. Herons, R. W., Mc Whirter, P., Rhoderick, E. H.: Proc. Roy. Soc. *A 234*, 565 (1956)
53. Bevington, P. R.: "Data Reduction and Error Analysis for the Physical Sciences" Mc Graw Hill, New York, 1969
54. "Synchrotron Radiation Research", H. Winnick, S. Doniach Ed., Plenum Press, New York, 1981
55. "Les applications du rayonnement synchrotron en Biologie à Lure" Coll. "Interfaces", éditions du C.N.R.S., Paris 1978
56. Wahl, P., Auchet, J. C., Donzel, B.: Rev. Sci. Instr. *45*, 28 (1974)
57. Rayner, D. M., Mc Kinnon, A. E., Szabo, A. G.: Can. J. Chem., *54*, 3246 (1976)
58. Sipp, B., Miehe, J. A., Lopez-Delgado, R.: Optics Comm. *16*, 202 (1976)
59. Valeur, B.: Chem. Phys. *30*, 85 (1978)
60. André, J. C., Lopez-Delgado, R., Lyke, R. L., Ware, W. R.: Appl. Optics, *18*, 1355 (1979)
61. Imhof, R. E., Birch, D. J. S.: Optics Comm., *42*, 83 (1982)
62. Rumbles, G.: in "Deconvolution and Reconcolution of Analytical Signals" M. Bouchy ed., ENSIC-INPL, Nancy (France), 1982
63. Haehnec, W., Nairn, J. A., Resiberg, P., Sauer, R.: Biochimica and Biophysica Acta, *680*, 161 (1982)
64. Wild, U., Holzwarth, A., Good, H. P.: Rev. Sci. Instr. *48*, 1621 (1977)
65. Wijnaendts van Resandt, R. W., de Maeyer, L.: Chem. Phys. Lett. *78*, 219 (1981)
66. Wijnaendts van Resandt, R. W., Vogel, R. H., Rovencher, S. W.: Rev. Sci. Instrum. *53*, 1392 (1982)
67. Rulliere, C., Declemy, A., Pee, P.: Rev. Phys. Appl. *18*, 39 (1982)
68. Durbin, J., Watson, G. S.: Biometrika, *37*, 409 (1950); *38*, 159 (1951)
69. Williams, G.: Adv. Pol. Sci., *33*, 59, (1979)
70. Patterson, G. D., Lindsey, C. P., Stevens, J. R.: J. Chem. Phys. *70*, 643 (1979)
71. Patterson, G. D., Lindsey, C. P.: Macromolecules, *14*, 83 (1981)
72. Patterson, G. D.: Adv. Polym. Sci., *48*, (1982)
73. de Gennes, P. G.: J. de Phys. (Paris), *42*, 735 (1981)
74. de Gennes, P. G.: "Scaling concepts in polymer physics", Cornell U. P., Ithaca, New York, 1979
75. Doi, M., Edwards, S. F.: J. Chem. Soc., Faraday Trans II, *74*, 1789; 1802; 1818 (1978)
76. Yeung, C. K.: Thesis, Paris VI, 1982

H. H. Kausch (Editor)
Received March 2, 1984

# Resonance Scattering in Macromolecular Structure Research

H. B. Stuhrmann
Institut für Physikalische Chemie, Universität Mainz, 6500 Mainz

*Resonance (or anomalous) X-ray scattering of partially ordered macromolecular structures, amorphous materials and solutions is encountered in the near edge region of X-ray absorption edges where the resonant real part f' of atomic form factors shows the strongest dispersion. The requirements of spectral brilliance in the near absorption edges can only be met by synchrotron radiation emitted from high energy electron (positron) storage rings. Resonance scattering yields three basic scattering functions. This compares to contrast variation in neutron scattering. The relations to isomorphous replacement methods of crystallography are discussed. The analysis of the basic scattering functions in terms of a multipole expansion provides a promising strategy, once the possibilities of synchrotron radiation are fully exploited so as to allow for the required high precision of the scattering measurements.*

**1 Introduction** . . . . . . . . . . . . . . . . . . . . . . . . . . 124

**2 Resonance Scattering** . . . . . . . . . . . . . . . . . . . . . 124
   2.1 X-rays and Neutrons . . . . . . . . . . . . . . . . . . . . . 126
   2.2 X-ray Absorption fine Structure . . . . . . . . . . . . . . . 130
   2.3 Scattering by Oriented Molecules . . . . . . . . . . . . . . 131
   2.4 Scattering by Free Molecules . . . . . . . . . . . . . . . . 135
   2.5 Scattering by Amorphous Material . . . . . . . . . . . . . 138

**3 Experimental** . . . . . . . . . . . . . . . . . . . . . . . . . 139
   3.1 Instruments . . . . . . . . . . . . . . . . . . . . . . . . . 140
      3.1.1 Double Monochromator Systems . . . . . . . . . . . . 140
      3.1.2 Focussung Mirror and Double Monochromator . . . . . 142
      3.1.3 Double Mirror System and Crystal Monochromator . . 143
   3.2 Absorption and Sample Dimensions . . . . . . . . . . . . . 144
   3.3 Estimates of Resonant Scattering . . . . . . . . . . . . . . 146
   3.4 Requirements for Data Aquisition . . . . . . . . . . . . . . 148

**4 Results** . . . . . . . . . . . . . . . . . . . . . . . . . . . . 149
   4.1 Ferritin . . . . . . . . . . . . . . . . . . . . . . . . . . . 149
   4.2 Parvalbumin . . . . . . . . . . . . . . . . . . . . . . . . 152
   4.3 Hemoglobin . . . . . . . . . . . . . . . . . . . . . . . . 153
   4.4 Membranes . . . . . . . . . . . . . . . . . . . . . . . . 155
   4.5 Polyelectrolytes . . . . . . . . . . . . . . . . . . . . . . 158
   4.6 Conducting Polymers . . . . . . . . . . . . . . . . . . . 160

**5 Conclusions** . . . . . . . . . . . . . . . . . . . . . . . . . 161

**6 References** . . . . . . . . . . . . . . . . . . . . . . . . . 162

# 1 Introduction

A new method in macromolecular structure research will have as many aspects and possibilities as there are different kinds of macromolecular self organisation. Too many to consider all of them in this article. A particularly simple case is that of a dilute solution of macromolecules. There the averaged structure factor can be measured either by light scattering or by small angle scattering of X-rays or neutrons. We shall refer to this case more often below, as the use of resonance scattering there presents a new and illuminating way of structure research. The advantages of resonance (or anomalous) scattering in crystallography were recognized much earlier when Bijvoet pointed out to small differences in Friedel related pairs of X-ray reflections. It is the aim of this article to guide the reader from well-established facts of resonance scattering to its application in new areas, where a succesful use of this technique only became possible with the advent of synchrotron radiation.

# 2 Resonance Scattering

Although the physical processes responsible for X-ray and neutron resonance scattering are vastly different a unified approach of resonance (or anomalous) scattering can be given on the basis of the famous 'optical theorem' [1].

The quantum theory of scattering also known as the method of partial wave analysis provides the frame work for this and we give here the main results of the theory. The basic step in this theory is the decomposition of the incident plane wave into various partial waves each with a well-defined angular momentum:

$$\exp(ik \cdot r) = \sum_{l=0}^{\infty} i^l (2l+1) j_l(kr) P_l(\cos\theta) \tag{1}$$

Here $\exp(ik \cdot r)$ represents the plane wave and the individual terms in the summation correspond to angular momentum eigen-functions of the Schrödinger wave equation with $V(r) = 0$. $j_l(kr)$, the spherical Bessel function, describes the spatial dependence of the l-th partial wave. The angular dependence of the l-th partial wave is given by the Legendre polynomial $P_l(\cos\theta)$. This description has considerable advantages. Firstly, in the presence of a spherically symmetric potential the effect of 'turning on' the potential on each of the partial waves can be considered separately. Secondly, if the range of the potential is small, the higher-angular-momentum partial waves are least affected by the presence of the potential.

The asymptotic form of the wave function in the presence of the potential is given by

$$\psi(r, \theta) = \exp(ikr \cos\theta) + \frac{b(\theta)}{r} \exp(ikr) \tag{2}$$

The first term in the right hand side represents the incident wave and the second term, the spherical wave diverging from the scattering centre. $b(\theta)$, which has units of length, is referred to as the 'scattering amplitude' or the 'scattering length' and is a central quantity in the theory of scattering. If the potential falls off more rapidly than $1/r^2$ then the asymptotic form of the wave equation [2] can be written as

$$\Psi(r, \theta) = \sum_{l=0}^{\infty} a_l R_l(kr) P_l(\cos \theta) \tag{3}$$

The function $R_l(kr)$ differs from $j_l(kr)$ in Eq. (1) as the boundary conditions near the scattering centre are different in the two cases. The asymptotic form of the spherical Bessel function $j_l(kr)$ is given by

$$j_l(kr) = \frac{\sin(kr - l\pi/2)}{kr} \quad r \to \infty \tag{4}$$

This represents a 'radial standing wave' and thus can be broken up into an incoming and an outgoing wave from the scattering centre.

For pure elastic scattering, the effect of the potential is just to alter the phase of the outgoing wave, i.e. $R_l(kr)$ will be different from $j_l(kr)$ in the asymptotic limit only through a phase factor.

$$R_l(kr) \to \frac{\sin(kr - l\pi/2 + \delta_l)}{kr} \quad r \to \infty \tag{5}$$

Here $\delta_l$ is the phase shift introduced in the l-th partial wave. Using the relations (5) and (4) in (2) one obtains the scattering amplitude

$$b(\theta) = \frac{1}{k} \sum_{l=0}^{\infty} (2l + 1) \sin \delta_l \exp(i\delta_l) P_l(\cos \theta) \tag{6}$$

In the case of inelastic processes, the asymptotic form of $R_l(kr)$ given in Eq. (5) no longer holds. This is because the amplitude of the outgoing spherical wave must necessarily be less than that of the incoming wave. These amplitudes however refer only to the elastic component of the scattering amplitude. The breakdown of the radial standing wave pattern in $R_l(kr)$ results in a nett inward flux towards the origin. In fact these are the waves which go into various inelastic channels. This is taken into account by

$$R_l(kr) = \frac{\eta_l e^{i(kr - l\pi/2)} - e^{-i(kr - l\pi/2)}}{2ikr} \tag{7}$$

where $\eta_l = A_l \exp(2i\delta_l)$, $A_l < 1$. Pure elastic scattering is a special case of this when $A_l = 1$. Using the asymptotic form of $R_l(kr)$ one gets for the elastic scattering amplitude the expression

$$b(\theta) = \frac{1}{2ik} \sum_{l=0}^{\infty} (2l + 1) [A_l \exp(2i\delta_l) - 1] P_l(\cos \theta) \tag{8}$$

It is clear that the scattering amplitude is basically a complex quantity. With

$$b(\theta) = b_0(\theta) + ib''(\theta)$$

we have

$$b_0(\theta) = \frac{1}{2k} \sum_l (2l + 1) A_l \sin 2\delta_l P_l(\cos \theta) \tag{9a}$$

and

$$b''(\theta) = \frac{1}{2k} \sum_l (2l + 1) [(1 - A_l) + 2A_l \sin^2 \delta_l] P_l(\cos \theta) \tag{9b}$$

The optical theorem relates the total cross section, which includes both elastic and inelastic contributions, to the imaginary component of the scattering amplitude in the exact forward direction. It is given by

$$\sigma_{total} = \sigma_{elastic} + \sigma_{inelastic} = 2\lambda b''(0) \tag{10}$$

Some important implications of the optical theorem may be summarized as follows:

1) Any scattering process, in general, should be characterized by a complex scattering amplitude. The absence of the imaginary component implies no scattering at all.

2) It is impossible to have only the inelastic scattering without elastic scattering, i.e. there cannot be absorption without elastic scattering.

3) The optical theorem provides a method of determining the imaginary component of the scattering amplitude in forward direction from the experimentally obtained total scattering scattering cross Section.

## 2.1 X-rays and Neutrons

When the wavelength of the incident X-ray beam is close to the absorption edge of an atom, the atomic scattering factor becomes complex to a greater extent

$$f = f_0 + f' + if'' \tag{11}$$

$f_0$ is the short wavelength limit of the scattering amplitude. Following the tradition of X-ray diffraction f is given in electrons, e.g. $f_0(0)$ of iron is 26. The scattering length of one electron is $2.8 \; 10^{-12}$ mm. $f'$ and $f''$ exhibit a wavelength dependence or dispersion. The absorption edge for X-rays represents the threshold frequency above which an inner electron can be ejected into the continuum. The resonance absorption also known as photoelectric absorption is an inelastic channel. The optical theorem for X-rays then is

$$\sigma_{photoelectric} = 2\lambda \frac{e^2}{mc^2} f''(0) \tag{12}$$

Since $\sigma_{photoelectric}$ exists only on the short wavelength region of the absorption edge, so also does f"(0). The dispersion of f"(0) is fully controlled by $\sigma_{photoelectric}$. For the K edge, $\sigma_{photoelectric}$ is represented fairly well by the empirical formula

$$\sigma_{photoelectric} = \left(\frac{\omega_K}{\omega}\right)^3 \sigma_{photoelectric}(\omega_K) \qquad \omega > \omega_K \qquad (13)$$

Since for photons k = ω/c, f" (0) should vary with the frequency roughly as $1/\omega^2$ Fig. 1 shows that the dispersion of f" (0) is slightly more complicated as ionization of p and d electrons will give rise to additional abrupt changes of f. In fact the schematic representation of the absorption edge in Fig. 1 has to be replaced by a more structured dispersion profiles at higher frequency resolution, as is shown in Fig. 2.

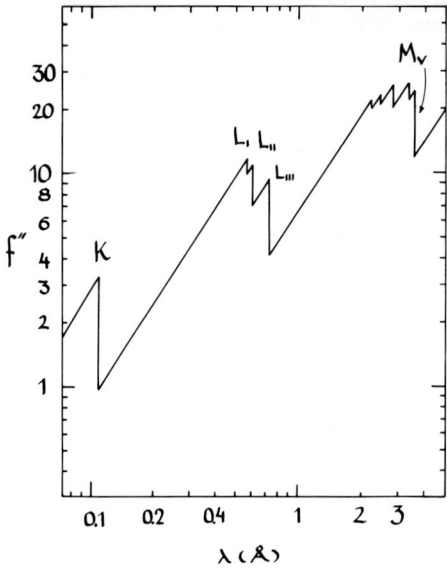

**Fig. 1.** Dispersion of the imaginary part f" of the atomic form factor of uranium. f" (in units of electrons) reaches about one third of the non-resonant atomic form factor f = 92. This schematic representation has been taken from "The International Tables of Crystallography, IV"

The absorption coefficient μ (in mm$^{-1}$) and density 1 Mg/m$^3$ is

$$\mu = \frac{2N_L}{M} \lambda \frac{e^2}{mc^2} f'' = \frac{337,1}{M} \lambda f'' \qquad (14)$$

($N_L$ = 6.02 10$^{23}$, M = atomic weight, $e^2/mc^2$ = 2.8 · 10$^{-12}$ mm)
X-ray absorption coefficients are listed in the 'International Tables of Crystallography III'[2]. From the numerous empirical formulae used for the Computation of X-ray absorption coefficents we just mention

$$\mu = 0.016 \lambda^3 Z^{3.94}/M \qquad \lambda < \lambda_K \qquad (15a)$$
$$\mu = 0.000529 \lambda^3 Z^{4.3}/M \quad \lambda > \lambda_K \qquad (15b)$$

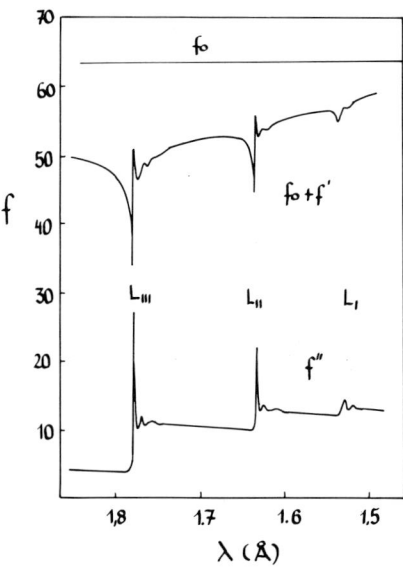

Fig. 2. Dispersion of $f_0 + f'$ of europium in Eu(PhAcAc)$_3$ at the three L-absorption edges after Lye, Phillips, Kaplan Doniachand Hodgson [4]). The f'-values were calculated by using the Kramers-Kronig relation (Eq. (17)). The absolute scale (in electrons) relies on values of f' and f'' at CrK$_\alpha$ and CuK$_\alpha$ which were taken from "The International Tables of Crystallography, IV"

Z is the atomic number. $\lambda_K$ is the wavelength at the K-absorption edge. On passing the X-ray K-absorption edge from lower to higher frequencies the absorption coefficient $\mu$ will increase by the factor $63.868/Z^{0.6207}$.

Since the spatial distribution of the core electrons is confined to a very small volume near the nucleus, f'' (and f') show a relatively weak dependence on the scattering angle. Only few experiments have been done to confirm theoretical predictions, e.g. on Barium with MoK$_\alpha$ radiation where

$$f'' = 2.40(5) - 0.59(11) \sin^2 \theta/\lambda^2 \quad [3)]$$

Any angular dependence of these quantities is due to the contributions from the p and higher-angular-momentum partial waves.

The physical process of resonance neutron scattering is through the formation of a 'compound nucleus'. $^{113}$Cd, $^{149}$Sm and $^{157}$Gd belong to the small class of nuclei which exhibit a resonance in the thermal energy region. In the case of $^{113}$Cd the compound nucleus $^{114}$Cd will either eject a neutron in an (n, n) process or emit $\gamma$-rays in an (n, $\gamma$) process the latter being inelastic. Unlike in X-ray anomalous dispersion, in the present case both the elastic (n, n) and inelastic (n, $\gamma$) processes contribute to b''(0):

$$\sigma_{(n,n)} + \sigma_{(n,\gamma)} = 2\lambda\, b''(0) \qquad (16)$$

The measurement of the total cross section determines completely the dispersion of f''. — The dispersion of the real part is not independent of that of f'', for there exists a general relationship between them known as the Kramers-Kronig relation:

$$f'(\omega) = \frac{2}{\pi} \int_{\omega'=0}^{\infty} \frac{\omega' f''(\omega')}{\omega^2 - \omega'^2}\, d\omega' \qquad (17)$$

Thus a knowledge of $f''(\omega)$ over a sufficiently wide frequency region permits the evaluation of $f'(\omega)$. With X-rays the frequency interval of strong anomalous dispersion may be very narrow. Then Eq. (14) simplifies to

$$f'(\omega) = -\frac{1}{\pi} \int_{\varepsilon=-\delta}^{\delta} \frac{f''(\omega+\varepsilon)}{\varepsilon} d\varepsilon \tag{18}$$

where $\delta$ is of the order 0.01.

The dispersion of the resonant scattering amplitudes of neutron scattering compare to those encountered at the $L_3$ absorption edges of rare earth ions. Fig. 3 shows however an important difference between these two kinds of resonance phenomena: nuclear resonance extends over a rather wide spectral range, whereas the region of strong dispersion is limited to the close vicinity at X-ray absorption edges. In fact, the intervall of dispersion may be so small that the angular distribution of X-ray reflections hardly changes in an resonance scattering experiment, whereas neutron

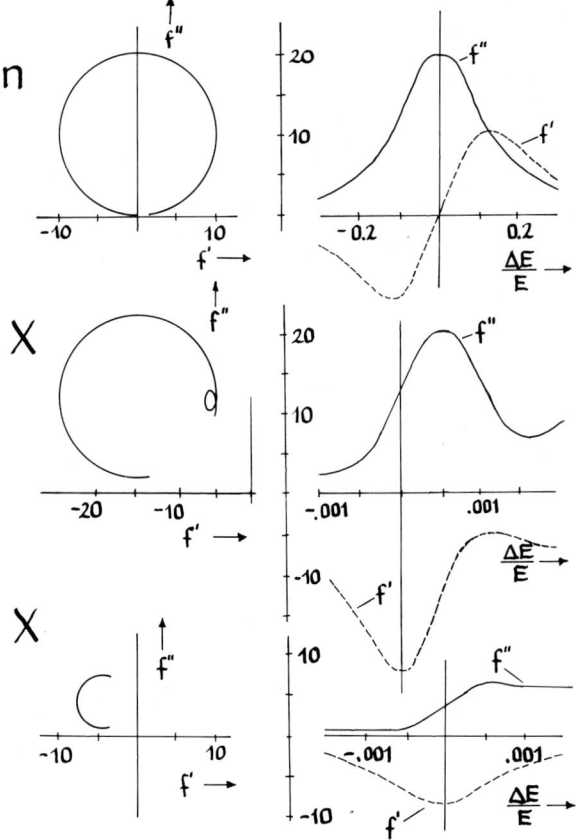

**Fig. 3.** Schematic representation of nuclear resonance scattering and X-ray resonance scattering at an K and an $L_3$ absorption edge. The phase digrams show the particular differences and similarities

diffraction patterns will move considerably in the detection plane on passing through the resonance. Resonant neutron scattering experiments may require isotopic enrichment of the nucleus in question, as the natural abundance of the resonant isotopes mentioned above hardly exceeds 15%.

The interplay of $b'$ and $b''$ becomes clearer when the amplitude b is represented in in the plane of complex numbers (Argand diagram). As is shown in Fig. 3b follows a circular line, which is nearly closed in the case of nuclear resonance. If the absorption edge resembles a step function, then the phase diagram will show a half circle. So called white lines at absorption edges form an intermediate class. They have gained considerable importance in X-ray resonant diffraction [4,5]. We also note slight deformations of the main circular line of X-ray phase diagrams and smaller epicyclic features which reflect the fine structure of the absorption edge. The amplitude b is nearly always positive with X-ray scattering, whereas it assumes both positive and negative values in neutron resonance scattering. This is so because $b_0$ of neutron scattering is about an order of magnitude smaller than the corresponding resonant terms.

## 2.2 X-ray Absorption Fine Structure

X-ray spectra [6] are distinguished from ordinary visible light or UV spectra by the fact that these photons can excite inner shell electrons from the absorbing atoms with consequent sharp steps in the absorption cross section as the X-ray energy is increased through an inner-shell ionization threshold. The resulting spectrum (Fig. 4) can be analyzed in two different regions [7,8].
— the threshold region from about 1 rydberg (Ry) below the absorption edge to 3–5 Ry above the absorption edge, and
— the extended X-ray absorption fine structure (EXAFS) or Kronig structure region from about 3–5 Ry above the edge out to as much as 100–150 Ry above the edge.

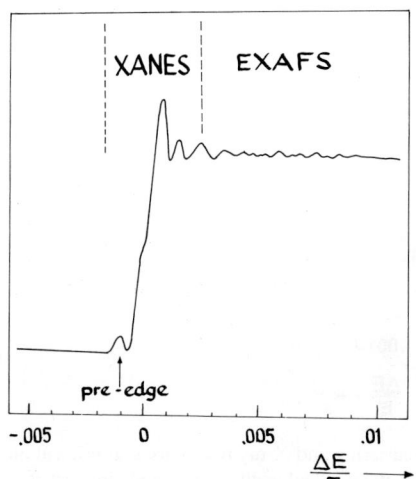

Fig. 4. The X-ray Absorption Near Edge Structure (XANES) and the Extended Absorption Fine Structure (EXAFS), — a schematic representation

In the edge region a photon has just sufficient energy to promote a core electron into the unoccupied atomic or molecular orbitals, i.e. discrete energy states or to low-lying continuum states, collective excitations, for instance plasmons. The energies and probabilities for these transitions depend on the formal charge of the absorbing atom, its actual charge, the symmetry of the site and the degree of ionicity and covalency of the bonds. Obviously, considerable information about the chemical environment is buried in the edge structures. The absolute positions of the absorption edge is also affected by the effects mentioned above. Traditionally it was assigned to its inflection point, a procedure that had to be revised.

At slightly higher energies, about 15 eV above the absorption edge the spectrum shows a series of gentle oscillations in the absorption cross section. These may be interpreted in a highly quantitative way in terms of the scattering of the excited photoelectrons by the neighbouring atoms and the resulting interference of this reflected electron wave with the outgoing photoelectron waves. An analogy could be made with a radar station which emits waves and detects the signals reflected from distant objects: neighbouring atoms reflect the outgoing electron wave back to the emitter and "detection" of the signal takes place through coherent interference of the returning wave with the emitted one [8]. This interference is sensitive to the amplitude and phase of the waves, i.e. to the number and distance of the surrounding objects. Tuning the photon excitation energy changes the wavenumber of the photoelectron wave as

$$k = \left[\frac{2m}{\hbar^2}(\hbar\omega - E_0)\right]^{1/2} \quad (19)$$

leading to oscillatory constructive and destructive interference between the emitted and backscattered waves. $E_0$ is the binding energy or ionization energy and consequently $\hbar\omega - E_0$ represents the kinetic energy of the photoelectron. These processes superimpose sinusoidal-like functions which have to be disentangled. Their periods give a direct measure of the distance between the emitter and scatterers, when appropriate corrections are made for the several phase shifts that take place [7,8]. This distance information extends out to about 3.5–4 Å from the absorbing species, and in favourable cases one may distinguish up to four different coordination distances in this bond-length range.

## 2.3 Scattering by Oriented Molecules

Now we turn to the right hand side of Equation (10) which describes the concomitant change of resonance scattering in f". The description of of resonance scattering starts from an assembly of M atoms each of them having fixed coordinates $r$ with respect to an origin. Out of these a smaller number N of atoms is supposed to have a strong resonance scattering. The structure $\varrho(r)$ is then represented by

$$\varrho(r) = \sum_{n=1}^{M} f_{0,n}\delta(r - r_n) + (f' + if'')\sum_{n=1}^{N} \delta(r - r_n)$$

$$= u(r) + (f' + if'')\,v(r) \quad (20)$$

It has been assumed that all strongly resonant scatters have the same dispersion i.e. they represent the same chemical element and they are assumed to have the same chemical environment (see Sect. 2.2). With $U(r)$ and $V(r)$ as the real and imaginary part of $\varrho(r)$ respectively, we obtain the scattering amplitude of the structure $\varrho(r)$ as

$$F(h) \sim \int [U(r) + iV(r)] e^{ihr} d^3r \qquad (21)$$

where $h$ is the scattering vector and $h = |h| = 4\pi \sin\theta/\lambda$. The multiplication of $F(h)$ with its complex conjugate

$$F^*(h) \sim \int [U(r) - iV(r)] e^{-ih\cdot r} d^3r \qquad (22)$$

yields the scattering intensity $S(h)$

$$\begin{aligned} S(h) &= A(h) A^*(h) \\ &\sim \int\int [U(r) + iV(r)] [U(r') - iV(r')] e^{ih\cdot(r-r')} d^3r\, d^3r' \\ &= \int\int [U(r) U(r') + V(r) V(r')] \cos[h\cdot(r-r')] d^3r\, d^3r' \\ &+ \int\int [U(r) V(r') - V(r) U(r')] \sin[h\cdot(r-r')] d^3r\, d^3r' \end{aligned} \qquad (23)$$

Separating the resonant real part we obtain

$$\begin{aligned} S(h) &\sim \int\int u(r) u(r') \cos[h(r-r')] d^3r\, d^3r' \\ &+ 2f' \int\int u(r) v(r') \cos[h(r-r')] d^3r\, d^3r' \\ &+ (f'^2 + f''^2) \int\int v(r) v(r') \cos[h(r-r')] d^3r\, d^3r' \\ &+ f'' \int\int [u(r) v(r') - v(r) u(r')] \sin[h(r-r')] d^3r\, d^3r' \\ &= S_u(h) + f' S_{uv}(h) + (f'^2 + f''^2) S_v(h) + f'' \varphi(h) \end{aligned} \qquad (24)$$

The overall effect of resonant scattering is to cause the break down of Friedel's law so that the Bijvoet pairs of reflections $S(h)$ and $S(-h)$ are unequal [9]. The difference

$$S(h) - S(-h) = 2 f'' \varphi(h) \qquad (25)$$

is used to determine the absolute configuration of a crystal structure [9]. The corresponding sine Patterson function has been discussed by Okaya, Saito and Pepinsky [10].

For a crystal assumed to be infinite in extent, $F(h)$ exists only at reciprocal lattice points. The structure factor Equation (21), written in terms of the indices of reflexion and fractional coordinates within the unit cell $\varrho(xyz)$ is

$$F_{hkl} = V \int_{x=0}^{1} \int_{y=0}^{1} \int_{z=0}^{1} u(xyz) + (f' + if'') v(xyz) e^{2\pi i(hx+ky+lz)} dx\, dy\, dz \qquad (26)$$

$V$ is the volume of the unit cell and should not be confused with $V(r) = V(xyz)$. From the fact that only the absolute square of F can be measured, the phase of the complex value of F is lost. This is the phase problem of crystallography, which found its solution by the introduction of the method of isomorphous replacement in the fifties [11].

To solve the phase problem for protein structure determination, one normally prepares a series of heavy-atom-containing crystals of the protein. They should all be identical to the crystals of protein with no heavy atom except for the inclusion of the extra atom. Such a crystal pair are called the native and the heavy atom isomorphous derivative. The difference between the native and derivative can be thought of as the diffraction pattern from a single atom. This can be solved for the position of the atom within the unit cell. Then the phase of the native structure may be estimated. Knowing the position of a heavy atom, the amplitude and phase of its contribution to the diffraction pattern can be calculated. Then the phase difference between the native and derivative can be calculated as one knows the amplitudes of two complex numbers and the amplitude and phase of the difference between them. There remains an ambiguity of phase that is resolved by using data from a second derivative. This method was originally demonstrated for protein structures by Perutz and co-workers [10]. In practice, a phase probability function is calculated and a Fourier transform with weighted coefficients is used [12]. Use of many derivatives improves the map.

An analogous procedure can be adopted by using resonance scattering. Protein molecules consist mostly of hydrogen, carbon, nitrogen, oxygen and sulphur atoms, which exhibit relatively weak anomalous dispersion. One heavier atom is supposed to show strong anomalous dispersion. The effect of the complex anomalous scattering term is shown in Fig. 5. The sum of the light atom scattering of the protein is shown as $F_p$ and the normal and anomalous scattering of the heavy atom are shown with the total measured scattering as their vector sum in the Argand diagram [13]. In the narrow wavelength range around an absorption edge only f' and f" will change while $F_p$ and $f_0$ stay constant. The vector of the total amplitude describes a circle as represented in Fig. 3. Each different wavelength corresponds to another point on this circle. This is equivalent to the scattering from different isomorphous derivatives. Also shown are the summations for the Bijvoet related reflection, with indices of opposite sign. For such reflections all real contributions are of the same magnitude but have a phase of opposite sign to the original reflection, that of f" retains the same phase. It is customary to represent this reflexion inverted through the real axis so that it lies totally on the original reflection except for the f" contribution, which then has the opposite sense. It can be seen that either shifting wavelength or measuring the Biojvoet related reflection produces an intensity change [14].

The measured intensity at a wavelength with anomalous dispersion terms f' and f" is given by

$$S = F_p^2 + (f_0 + f')^2 + f''^2 + 2F_p(f_0 + f') \cos \varphi + 2F_0 f'' \sin \varphi \quad (27)$$

where $\varphi$ is the phase angle between the protein and the heavy atom vectors. The sign is set according to which of the Biijvoet pair is being measured. If the anomalous dispersion terms are known, two measurements are enough to give the phase to within a sign ambiguity. A third measurement will resolve this ambiguity, provided the three vector sums in the phase diagram do not fall on a straight line. Multi-wavelength methods are gaining considerable interest in recent resonance diffraction work [15].

A very elegant approach for the solution of the phase problem in the terms of Eq. (24) has been suggested by K Fischer [16]. If three wavelengths near an absorption edge are chosen in such a way that $f'_{\lambda 1} = f'_{\lambda 2} \neq f'_{\lambda 3}$ and $f''_{\lambda 1} \neq f''_{\lambda 2} = f''_{\lambda 3}$ then the Fouriertransform of the differences in the intensities measured at $\lambda_1, \lambda_2$ and $\lambda_2, \lambda_3$ respectively separates the convolution of the resonant atoms with the other atoms, while the convolution of the nonresonant atoms vanishes (Lamda-method).

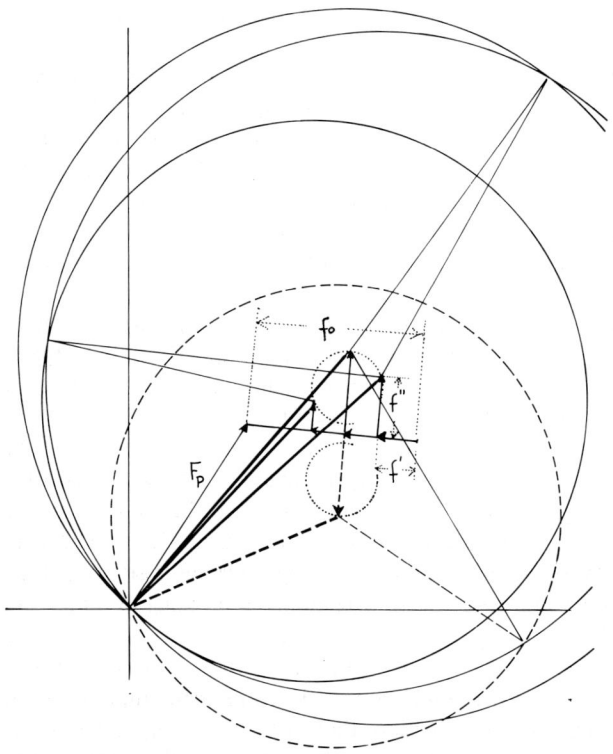

**Fig. 5.** The amplitude and phase of a Bragg reflection from a protein crystal containing a heavy atom is represented as a vector on the complex plane, after Phillips and Hodgson [12]. F is the vector sum of the protein contribution $F_p$ and $f_0$ is the normal scattering of the heavy atom; the other abbreviations are given in the text

Once the phases of all $F_{hkl}$ have been determined the structure $\varrho(r)$ can be determined by Fourier synthesis.

$$\varrho(xyz) = \frac{1}{V} \sum_{h=-\infty}^{\infty} \sum_{k=-\infty}^{\infty} \sum_{l=-\infty}^{\infty} F_{hkl}\, e^{-2\pi i (hx + ky + lz)} \tag{28}$$

The resulting scattering density map (electron density in the case of X-ray diffraction) usually is processed in further refinement methods in order to obtain a molecular model at atomic resolution [9].

## 2.4 Scattering by Free Molecules

A first step towards the consideration of free, randomly oriented molecules (or microcrystals) is the superposition of S(h) with S(−h):

$$\frac{1}{2}[S(h) + S(-h)] = S_u(h) + f' S_{uv}(h) + (f'^2 + f''^2) S_v(h) \tag{29}$$

The dispersion of the cosine term in Eq. (25) has the same form. Further averaging processes as they occur with scattering from randomly oriented macromolecules in solution do not change the dispersion given in Eq. (29). Introducing the wavelength dependence of f' and f'' explicitely, we obtain by integration of S(h) over the solid angle $\Omega$ [17]

$$\begin{aligned} I(h, \lambda) &= \int S(h, \lambda) \, d\Omega \\ &= \int\int u(r)\,u(r')\sin h|r-r'|/|r-r'|\,d^3r\,d^3r' \\ &\quad + 2\,f'(\lambda) \int\int u(r)\,v(r')\sin h|r-r'|/|r-r'|\,d^3r\,d^3r' \\ &\quad + [f'^2(\lambda) + f''^2(\lambda)] \int\int v(r)\,v(r')\sin h|r-r'|/|r-r'|\,d^3r\,d^3r' \\ &= I_u(h) + f'(\lambda)\,I_{uv}(h) + [f'^2(\lambda) + f''^2(\lambda)]\,S_v(h) \end{aligned} \tag{30}$$

Macromolecules in solution as descibed by Eq. (20) give rise to three basic scattering functions. $I_u(h)$ is the off-resonance scattering, $I_v$ originates from the structure of the resonant scatterers alone, and $I_{uv}(h)$ is a cross term, containing the influence of both u(r) and v(r) as their convolution. In many cases the resonant scattering terms in Eq. (30) may be much smaller than $I_u(h)$. The quadratic term in Eq. (30) then can be neglected and we obtain

$$I(h, \lambda) = I_u(h) + f'(\lambda)\,I_{uv}(h) \tag{31}$$

This equation stresses the dominant role of f' in resonant scattering of noncrystalline structures. As can be deduced from Fig. 3 resonant scattering in this case will be observed in the very close vicinity of the absorption edge. This is quite different from diffraction by non-centrosymmetric unit cells where the f'' dispersion according to Eq. (25) allows resonance scattering measurements even at much shorter wavelengths than that of the absorption edge [18]. The tunability of synchrotron radiation to the narrow regions of strong dispersions of f' near the absorption edge for the first time created the necessary conditions for successful measurements of resonance scattering from disordered systems which is basically linked to $f'(\lambda)\,I_{uv}(h)$. This point will be discussed in more detail below.

The analysis of the resonant solution scattering data demands a different representation of the Debye Equation (30). If the macromolecular structure would have a spherical appearance, then the formalism of isomorphous replacement in single crystal diffraction outlined in the preceding section would apply. This is not surprising as the rotation of a spherical structure could not be noticed anyhow. In more complicated, asymmetric macromolecular structures it is the spherical average of the structure which can be subjected to the phase analysis described above. As this statement is less trivial, we shall extend the description of a macromolecular structure beyond its spherical average by introducing an expansion of $\varrho(r)$ as a series of spherical harmonics $Y_{lm}(\omega)$

$$\varrho(r) = \sum_{l=0}^{\infty} \sum_{m=-l}^{l} \varrho_{lm}(r) Y_{lm}(\omega) \qquad (32)$$

where $\varrho_{lm}(r) = \int \varrho(r) Y_{lm}^*(\omega) d\omega$

$\omega$ is a unit vector in physical space.

The amplitude $F(h)$ again is represented as a sum of partial waves (compare Sect. 2.1)

$$F(h) = \sum_{l=0}^{\infty} \sum_{m=-l}^{l} F_{lm}(h) Y_{lm}(\omega) \qquad (33)$$

where the radial functions are uniquely related by Hankel transformations

$$F_{lm}(h) = \sqrt{\frac{2}{\pi}} i^l \int \varrho_{lm}(r) j_l(hr) r^2 dr \qquad (34)$$

$$\varrho_{lm}(r) = \sqrt{\frac{2}{\pi}} (-i)^l \int F_{lm}(h) j_l(hr) h^2 dh \qquad (35)$$

On averaging $F(h) F^*(h)$ with respect to all orientations ($\Omega$) in reciprocal (or momentum) space we obtain [19]

$$I(h) = 2\pi^2 \sum_{l=0}^{\infty} \sum_{m=-l}^{l} F_{lm}(h) F_{lm}^*(h) \qquad (36)$$

This is the form of a scalar product which can be written as $I = \langle F|F \rangle$. Each multipole $\varrho_{lm}(r) Y_{lm}(\omega)$ has its own scattering function $|F_{lm}(h)|^2$. In particular, the scattering of the average structure appears as the monopole scattering $|F_\infty(h)|^2$.

This representation of the scattering from randomly oriented particles would be of nearly no importance, unless there would not be a general way of separating the various multipole contributions to $I(h)$. In fact, Svergun, Feigin and Schedrin have proposed an algorithm which allows to distinguish between groups of multipoles with the same index l [19].

$$\sum_{m=-l}^{l} \varrho_{lm}(r) Y_{lm}(\omega) \leftrightarrow \sum_{m=-l}^{l} |F_{lm}(h)|^2 \qquad (37)$$

This means that there is a way to seperate the monopole scattering ($l = 1$) from the scattering of the three dipoles $p_x$, $p_y$, $p_z$ ($l = 1$; $m = -1, 0, 1$) and the five quadrupole functions ($l = 2$; $m = -2, -1, 0, 1, 2$) and increasingly bigger groups of higher multipoles [20]. To merge this mathematical analysis of $I(h)$ with the method of resonance scattering we modify Eq. (32)

$$\varrho(r) = \sum_{l=0}^{\infty} \sum_{m=-l}^{l} [u_{lm}(r) + (f' + if'') v_{lm}(r)] Y_{lm}(\omega) \qquad (38)$$

$$F(h) = \sum_{l=0}^{\infty} \sum_{m=-l}^{l} [A_{lm}(h) + (f' + if'') B_{lm}(h)] Y_{lm}(\Omega) \qquad (39)$$

The dispersion of I(h) then writes

$$I(h) = \sum_{l=0}^{\infty} \sum_{m=-l}^{l} A_{lm}^2(h) + f'[A_{lm}(h) B_{lm}^*(h) + B_{lm}(h) A_{lm}^*(h)]$$

$$+ (f'^2 + f''^2) B_{lm}^2(h). \qquad (40)$$

More explicitly and omitting (h) we obtain

$$
\begin{aligned}
I(h) = A_{00}A_{00} & \quad + 2f' A_{00}B_{00} & (f'^2 + f''^2) B_{00}B_{00} \\
A_{10}A_{10} & \quad + 2f' A_{10}B_{10} & (f'^2 + f''^2) B_{10}B_{10} \\
A_{1-1}A_{1-1}^* & \quad f'(A_{1-1}B_{1-1}^* + B_{1-1}A_{1-1}^*) & (f'^2 + f''^2) B_{1-1}B_{1-1}^* \\
A_{11}A_{11}^* & \quad f'(A_{11}B_{11}^* + B_{11}A_{11}^*) & (f'^2 + f''^2) B_{11}B_{11}^* \\
A_{20}A_{20} & \quad + 2f' A_{20}B_{20} & (f'^2 + f''^2) B_{20}B_{20} \\
A_{2-1}A_{2-1}^* & + f'(A_{2-1}B_{2-1}^* + B_{2-1}A_{2-1}^*) & + (f'^2 + f''^2) B_{2-1}B_{2-1}^* \\
A_{21}A_{21}^* & + f'(A_{21}B_{21}^* + B_{21}A_{21}^*) & + (f'^2 + f''^2) B_{21}B_{21}^* \\
A_{2-2}A_{2-2}^* & + f'(A_{2-2}B_{2-2}^* + B_{2-2}A_{2-2}^*) & + (f'^2 + f''^2) B_{2-2}B_{2-2}^* \\
A_{22}A_{22}^* & + f'(A_{22}B_{22}^* + B_{22}A_{22}^*) & + (f'^2 + f''^2) B_{22}B_{22}^*
\end{aligned}
$$

more blocks with 2l + 1 lines may follow (41)

The strategy of data analysis is now at hand. Contrast variation determines the columns of Eq. (41) whereas groups of rows of 2l ∗ 1 terms can be identified with the Svergun method [20].

For l = 0 the monopole structure can be determined completely if $B_{00}(h)$ is known. In fact, the phase problem of $F_{00}^2$ reduces to the determination of the sign of $F_{00}$. This is usually not too difficult a task for $B_{00}$ as plausible arguments can be made concerning the corresponding radial mass distribution $v_{00}(r)$ of the resonant label atoms. Once the signs of the sinusoidal function $B_{00}(h)$ are known for each peak, the phases of $A_{00}(h)$ can be determined directly by using the cross term.

If the structure is elongated the quadrupole terms (d-functions) have to be considered. Out of the five terms three can be elimated by rotation of the structure by the Eulerian angles α, β and γ. Thus we are left with [21]

$$I2_u(h) = A2_1(h) A2_1(h) + A2_2(h) A2_2(h) \qquad (42)$$

$$I2_{uv}(h) = A2_1(h) B2_1(h) + A2_2(h) B2_2(h) \qquad (43)$$

$$I2_v(h) = B2_1(h) B2_1(h) + B2_2(h) B2_2(h) \qquad (44)$$

I2 denotes the basic scattering functions for l = 2. It is also assumed that the complex radial functions have been converted to real functions A2 and B2 by appropriate linear combinations of the $A_{2m}$ and $B_{2m}$. Again it is assumed that the spatial distribution of the resonant atoms is simple enough to guess the signs of $B2_1(h)$ and $B2_2(h)$. Then $A2_1(h)$ and $A2_2(h)$ can be found by a simple geometrical construction which has to be made for each h-Intervall. In the $A2_1$–$A2_2$-plane Eq. (42) represents a circle which is intersected by a straight line defined by Eq. (43). There will be two pairs $(A2_1, A2_2)$ as solutions of the Eq. (42–44). By using an approximation of the radial functions by rather few members of a series of polynomials a correlation between the solution in various h-intervals can be achieved. Laguerre polynomials are of particular interest because of their simple transformation properties [19,20].

The evalution of the three $A_{1m}$ leads to a construction in three-dimensionals space. The solutions are found as the intersection line of a plane with the surface of a sphere. With increasing index l of the multipoles the correlation between the resonant structure and the whole structure through the basic scattering function $I_{uv}(h)$ usually gets weaker and weaker.

As an important result of this joint use of resonant scattering and advanced multipole analysis we note that partial structure $\sum_m u_{lm}(r) Y_{lm}(\omega)$ can be split into its constituents by the introduction of the known resonant label structures $v_{lm}(r) Y_{lm}(\omega)$. In the case of the quadrupole structure, one obtains the relative orientation of the main axes between the resonant structure and the total molecule. If a different resonant structure can be used, more information can be obtained, as the non-resonant structure gets convoluted with an other reference structure as a probe. This procedure might gain considerable importance in ribosome structure work [22].

Very often it is not possible to obtain all basic scattering functions with the same accuracy. Even worse, in resonant scattering we are often left with the cross term $I_{uv}(h)$ only. This is still quite an acceptable situation, if the resolution to a monopole approximation of the structure is required, as $A_{00}(h)$ and $B_{00}(h)$ may be determined completely from the two remaining functions $I_u(h)$ and $I_{uv}(h)$. However a straight-forward method for the evaluation of higher multipoles in the sense of the above calculation then does not exist any more. The analysis of resonant scattering in this case has to refer to models.

## 2.5 Scattering by Amorphous Material

Due to the lack of long range periodicity, the structure of an amorphous material cannot be described as completely as that of a crystalline material. The dense packing of molecules introduces some short range regularity, which is described in terms of distribution functions. These give the average probability of finding an atom or a molecule in a specified volume element at some distance and direction from another atom or set of atoms. A particularly useful distribution function is the partial distribution function (PDF) that relates the probability of finding a particular pair of atoms at a particular interatomic separation. The partial distribution functions (or, to abbreviate further, pair functions) $\varrho_{uv}(r)$ can be mathematically defined as

$$4\pi r^2 \varrho_{uv}(r) = \frac{1}{N_u} \sum_{i=1}^{N_u} \sum_{j=1}^{N_v} \delta(|R_{mn}| - r) \tag{45}$$

Here $N_u$ is the number of atoms of the species u in the sample and $R_{ij}$ is the vector connecting atoms m and n. Sums over m and n will be over all atoms of the species in a sample, while sums over u and v will be over the species in the sample. Since an amorphous material lacks translational symmetry the probability of finding an pair of atoms or molecules at a given separation rapidly approaches the average probability as the separation increases.

$$4\pi r^2 \varrho_{uv}(r) \to 4\pi r^2 \varrho^v_0 \quad \text{as } r \to \infty \tag{46}$$

where $4\pi r^2 \varrho^v_0$ is the average density of species v.

In an elemental material there is only one pair distribution function. As the complexity of the system increases, the number of distribution functions rapidly increases. In a binary mixture there are three pair correlation functions and in a mixture with N components there are $(N + 1)N/2$.

To relate these pair functions to the X-ray scattering, we rewrite the Debye Equation (20) by substituting Eq. (46) as:

$$I(h) = \sum_u \sum_v N_u f_u(h) f_v(h) \int \varrho_{uv}(r) \sin(hr)/(hr) \, dV \tag{47}$$

where f(h) are atomic or spherical molecular form factors. Eq. (47) can be separated into terms due to the average density and deviations from the average density yielding [23]:

$$I(h) = \sum_u \sum_v x_u f_u(h) f_v(h) I_{uv}(h) \tag{48}$$

Here $x_u$ is the atomic fraction of species u. For a binary material the intensity is given by a linear combination of three partial structure factors weighted by the atomic scattering factors. Keating [24] has proposed that three independent intensities be collected by modifying the atomic scattering factors. A set of three independent equations is produced which can then be solved for the partial structure factors. Explicitely, these equations can be written (for a specific value of h) as:

$$I(E_i) = f\, x_u S_{uu} + 2f'(E_i) f_u x_u S_{uv} + [f'^2(E_i) + f''^2(E_i)] x_v S_{vv} \tag{49}$$

where $E_i$ is the photon energy associated with the i-th measurement.

## 3 Experimental

The measurement of resonance scattering from disordered systems requires scattering experiments at wavelengths in the very near vicinity of the absorption edge, i.e. in the threshold region of absorption or XANES region. The tunability of the X-ray source to any wavelength is a crucial demand. A continous X-ray spectrum fulfils this requirement. X-ray tubes emit to an appreciable extent monochromatic radiation as a consequence of the decay of inner-shell excitations. For instance a copper anode may emit roughly about half of its radiation power as $K_\alpha$-radiation which has a wavelength of 1.54 Å. The spectral density at other wavelengths is about a hundred times lower. Tungsten tubes are doing slightly better with respect to the production of "white radiation". Synchrotron radiation is magnetic bremsstrahlung emitted by highly relativistic electrons circulating on macroscopic orbits with dimensions comparable to a football plane. Its spectral brilliance exceeds that of X-ray tubes by about five orders of magnitude [6, 7]. Even rotating anodes are 10 000 times inferior to electron storage rings. This explains why diffuse resonant scattering techniques could not develope until X-ray synchrotron radiation became available. As we shall see later the high intensity of synchrotron radiation is absolutely needed. The small contribution of resonance scattering to the total scattering amounts to typically 0.001 to 0.05 and this fraction tends to be obscured by the much larger dispersion of the absorption and fluorescence (see optical theorem, Eq. (20)).

An instrument for resonance scattering has to fulfil two requirements:
— easy and highly reproducible selection of narrow band wavelength intervalls
— high accuracy and resolution of the scattering intensity pattern

The hybridization of a rapidly tunable spectrometer with a diffractometer has been achieved to various extents. There are high-performance spectrometers for the measurement of the fine structure of X-ray absorption edges on the one side and commercial diffractometers for crystallography on the other side. The existing instruments for resonance scattering experiments are a modular arrangement of both. More recent designs are based on a stronger mutual relation between the monochromator and the diffractometer,

## 3.1 Instruments

The existing instruments for resonance scattering experiments may be classified according to the monochromator system. As synchrotron radiation is highly polarized in the plane of the orbit, the vertical reflection by the monochromator crystals is preferred. Under these conditions the polarization factor remains nearly constant. With reference to this design feature, the following arrangements exist or are under construction:
— double monochromator
— focusing toroidal mirror, double monochromator
— focusing toroidal mirror, exchangeable plane mirror, crystal monochromator

### 3.1.1 Double Monochromator Systems

The most common type of double monochromators are built as channel-cut crystals. These are single crystals of silicon or germanium with a grove cut parallel or asymmetric to the 111 plane or 220 plane [7, 8]. The incident beam is reflected twice under exactly equal glancing angles and thus maintains its forward direction. The distance between the reflecting planes is only a few millimeters. The vertical beam movement due to rotation during one wavelength scan is a small fraction of a millimeter. More recently built double monochromator systems allow for an independent rotation of the two crystal faces and displacement of the second axis of rotation to maintain a constant position of the monochromatic beam.

An early type of this double monochromator is used with the instrument X15 at EMBL, Hamburg [25]. This consists of two separate single crystals. The first crystal is 24 m from the point of emission of synchrotron radiation. The second crystal is vertically displaced by 1.22 m. It can be moved along an inverted optical bench of 3 m length, resulting in vertical deflection angles 2θ from 18° to 60° (Fig. 6). The beam emerging from the second crystal is parallel with respect to the main beam and has a constant position of 1.22 m at any glancing angle θ of the two monochromator crystals. The wavelengths reflected by the monochromator crystals are given by Bragg's equation

$$n\lambda = 2d \sin \theta \qquad (50)$$

d is the spacing of the reflection planes, e.g. 2d = 6.53 Å for Ge (111) and 2d = 3.84 Å for Si (220). With 9° < θ < 30°, the wavelength range extends from λ = 0.6 Å with Si (220) at 9° to λ = 3.25 Å with Ge (111) at θ = 30°. This means that the K-absorption edges of elements from Z = 20 (Ca) to Z = 42 (Mo) and $L_3$ absorption edges from Z = 50 (Sn) onwards can be used for anomalous scattering experiments.

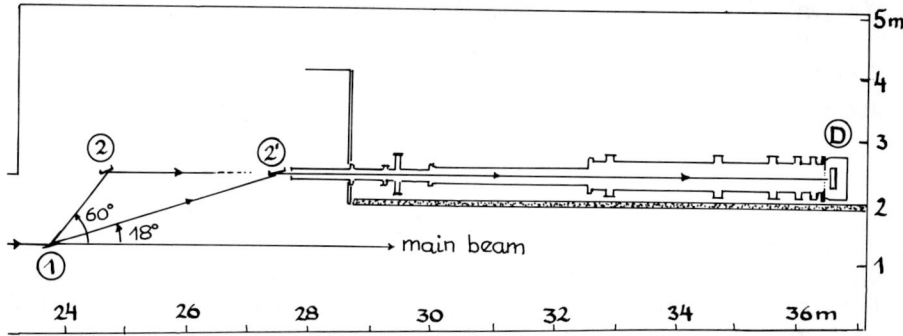

**Fig. 6.** The instrument X15 of EMBL at the storage ring DORIS, DESY Hamburg. The double monochromator system extends 24 to 28 m from the source. The first monochromator crystal (1) deflects the beam to the second crystal which can move between the positibns marked by (2) and (2'). The sample can be put at different distances from the detector (D)

The transmission function for a perfect crystal is strictly calculable from the dynamical theory [26], and if good crystals are used the experimental results follow closely the theoretical predictions. The choice of the perfect crystal is particularly important if one uses a double reflection geometry, e.g. a channel cut, because in this case, if imperfect crystals are used, not only will the wavelength resolution be poor, but large losses in flux occur. On the other hand if the crystal is perfect the use of a double reflection geometry leads to a marked improvement in the ratio of the peak to tail intensity, since within the confines of the rocking curve the reflectivity is almost 100% at wavelengths shorter than 2 Å. Fig. 7 shows a characteristic profile for a silicon 220 reflection at a wavelength of 1.54 Å.

Substitution of the acceptance angle due of the crystal in the differential form of Bragg' a law leads to the following formula for the energy resolution of the crystals:

$$\frac{\Delta E}{E} = \frac{\Delta \lambda}{\lambda} = \cot \theta \, \Delta\theta_{geom}. \qquad (51)$$

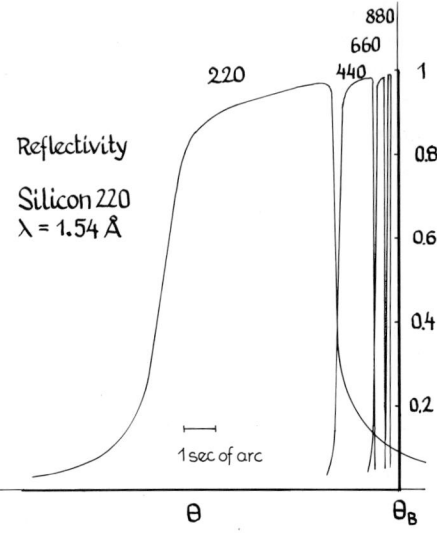

Fig. 7. The reflection characteristics of a silicon crystal 220 reflection at a wavelength of 1.54 Å. Observe the departure from the simple law of Bragg due to refractive index corrections

It can be shown that [27]

$$\frac{\Delta\lambda}{\lambda} = d_{hkl}^2 F_{hkl} \tag{52}$$

where $d_{hkl}$ is the spacing of the planes in use and $F_{hkl}$ is the structure factor for the reflection. If the structure factor is constant, i.e. away from an absorption edge, then the formula above tells us that the energy resolution of a crystal is independent of the wavelength. It is about $6 \cdot 10^{-5}$ for silicon (220) and about twice as much for Ge (111). Although from the point of view of monochromators very high energy resolution is possible, there is always the limitation on the resolution attainable due to the size of the source, the size of the slits to collimate the beam and the distance between the monochromator and the source. A working formula which gives the resolution of a monochromator in a practical situation is

$$\frac{\Delta\lambda}{\lambda} = [(d_{hkl}^2 F_{hkl}) + (\cot(\theta)\,\Delta\theta_{geom})]^{1/2} \tag{53}$$

where $\Delta\theta_{geom}$ is the angular aperture allowed by the slit system and source size. In practice the wavelength resolution of double crystal monochromators used for spectroscopic experiments is about $10^{-4}$.

### 3.1.2 Focussing Mirror and Double Monochromator

Once the the energy resolution requirements are established the vertical collimation of a synchrotron radiation beam is defined. This is not so with the horizontal divergence of the beam, where focusing elements in the beam path can increase the intensity of the sample considerably with nearly no deterioration of the wavelength resolution. One method to achieve focusing relies on the use of total reflection of X-ray mirrors. The refractive index for X-rays is given by

$$n = 1 - \frac{Ne^2\lambda^2}{2\pi mc^2} \tag{54}$$

where N is the electron density, $e^2/mc^2 = 2.8 \cdot 10^{-12}$ mm is the scattering length of an electron. For a given wayelength total reflection will occur up to a critical angle $\theta_c$:

$$\theta_c = \lambda \left(\frac{Ne^2}{\pi mc^2}\right)^{1/2} \tag{55}$$

This tells us also that for a given angle $\theta$ only wavelengths down to definte short wavelength limit will be reflected. X-ray mirrors are therefore often used to cut off the short wavelength part of the spectrum in order to ensure the absence of higher harmonics ($n \geq 2$ in Eq. (50)) on reflection by a crystal. Often quartz mirrors are used. Then $\theta_c = 0.00263\ \lambda$ (Å). With gold the critical angle is about 2.4 times higher.

In practice, the sharp cut-off predicted above never occurs, because absorption processes are not totally negligible. A sizable absorption means that the refractive index cannot be regarded as a real number and its complex nature (compare Eq. (11)) leads to a dispersion in the optical constants which results in a broadening of the sharp cut-off.

The total reflection of mirrors can be used to focus the radiation. Synchrotron radiation, while collimated in the vertical plane it spreads over the horizontal one. Wavelength resolution requirements normally restrict the vertical aperture to one mm or so, in any case. On the other hand it is desirable to condense the horizontal spread into a focal point. Double focussing with a mirror system is possible and an ideal mirror geometry has been worked out [28]. For a point source the mirror has to be shaped like an ellipsoid, and the source and the image have to be placed in the respective foci. The long distances involved in synchrotron work mean that a good approximation to shape is achieved by making use of bent cylindrical mirrors [7]

### 3.1.3 Two Mirrors, one Crystal Monochromator

This design is shown in Fig. (8). It has come up with the forthcoming use of resonant scattering in the soft X-ray region [29]. At wavelengths longer than 2 Å the maximum reflectivity more and more falls below 100% reflectivity. With Ge(111) at a wavelength of 6 Å the reflectivity reaches no more than 15% [30]. A double monochromator would transmit at most $0.15 * 0.15 = 2.5\%$ of the incident radiation. The use of only one crystal monochromator minimizes this loss. The price to be paid is the vertical rotation of the diffractometer around the monochromator axis.

The two mirrors have a different task. The first toroidal mirror images the source in a 1:1 ratio onto the site of the diffractometer. It also acts as a premonochromator. At a mean glancing angle of 0.007 the gold coating of the mirror will reflect wavelengths which are longer than 1 Å. This mirror has a length of 1 m and a

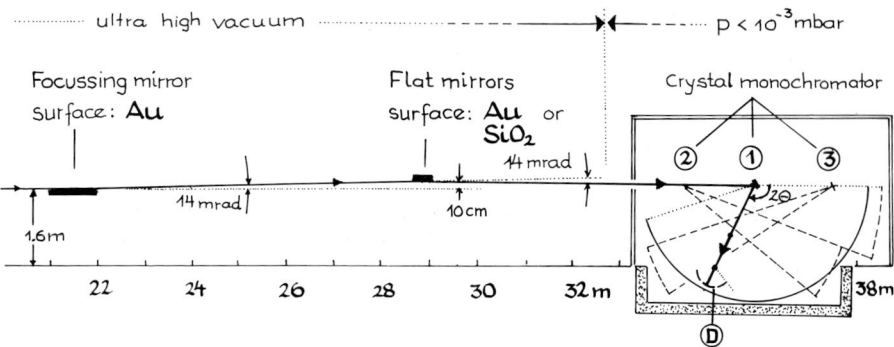

**Fig. 8.** The planned instrument for resonance scattering in the normal and soft X-ray spectrum of synchrotron radiation. It is presently under construction and will be installed at the beam line A1 of HASYLAB. The beam path is kept constant till the crystal monochromator. The camera is rotated by 2 θ in the vertical plane in order to follow the θ movement of the monochromator crystal, during an energy scan. The camera is evacuated. There is only a very thin window between the second mirror and the monochromator

horizontal width of 8 cm. The second, plane mirror acts as a further premonochromator. Reflecting the beam downwards by the same glancing angle $\theta = 0.007$ as above, this relatively small mirror will fulfil two pourposes:
— it establishes the original direction of the beam with a vertical displacement of 12 cm.
— it allows a change of the cut-off wavelength by lateral displacement of the mirror so that the uncoated quartz side gets exposed to the beam instead of the gold coated side. With the constant glacing angles given above, the cut-off wavelength changes from 1 Å to about 2.5 Å.

This is of utmost importance for the operation of the crystal monochromator. Using the 111-plane of crystals with a diamond lattice (e.g. Si, Ge, InSb, ...) the second harmonic (n = 2 in Eq. (50)) is suppressed, so that only wavelengths as short as $\lambda/3$, $\lambda/4$ would get reflected as well. The above mirror system in its gold — gold configuration would allow the transmission of wavelengths longer than 1 Å. The crystal then produces a monochromatic beam up to wavelengths of nearly 3 Å. Beyond this limit the crystal would find $\lambda/3$ as well unless one changes to the gold-quartz configuration on the mirror system. Then the X-ray spectrum incident on the crystal monochromator contains only wavelengths longer than 2.5 Å, and a monochromatic beam free of higher harmonics up to wavelengths of 7.5 Å can be produced. This limit matches the maximum wavelength of 7.48 Å transmitted by the 111-plane of an InSb single crystal. — An instrument of this type is being built at HASYLAB. It will allow X-ray resonance scattering studies with lighter elements like sulfur, phosphorus and silicon as well.

## 3.2 Absorption and Sample Dimensions

The absorption by a sample with the thickness d is given by

$$\frac{I}{I_0} = e^{-\mu d} \tag{56}$$

The optimal thickness of the sample for maximum scattering intensity in a transmission experiment is $1/\mu$. As the scattering intensity follows $d/\exp(\mu d)$ 82% of the maximum scattering intensity are reached at half the optimal thickness defined above. There may be two reasons which favour thin samples:
— the limited amount of material, a case which occurs often in biological applications
— the reduction of the dispersion due to absorption, which is very often necessary as the dispersion of resonant scattering is notoriously weak.

The relative change of the intensity due to anomalous dispersion of the absorption is

$$\frac{\Delta I}{I} = \frac{e^{-(\mu_0 + \mu_a)d} - e^{-\mu_0 d}}{e^{-\mu_0 d}} \approx \mu_a d \tag{57}$$

where $\mu_a$ is the anomalous dispersion of absorption and $\mu_0$ hardly changes near an absorption edge. There are two extreme cases, which lead to a different use of anomalous dispersion:

1. The concentration of anomalous scatterers is very high, i.e. more than 10%. Then resonant scattering can be measured with reasonable precision only on the long-wavelength-side of the absorption edge, where the dispersion of f' shows a relatively strong variation compared to the nearly constant absorption, as is shown in Fig. 2 [23]. Further decrease of the sample will not allow to use resonant scattering in the absorption edge as the onset of strong fluorescence will obscure resonant scattering contributions (see Sect. 4.5).

2. Lower concentration of anomalous scatterers, especially if they exhibit some regular structure, allow the measurement of the complete resonant dispersion profiles. The case of very dilute solutions of resonant scatterers will be discussed in Section 3.4.

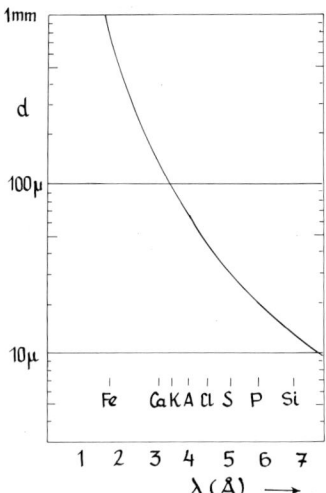

Fig. 9. The optimal thickness of a water sample in X-ray scattering experiments

As a rule of thumb, the dispersion of absorption and fluorescence should exceed resonant scattering by not more than one order of magnitude. This can be achieved by an appropriate choice of the sample thickness and the concentration of the resonant scatterers, as will be shown in Section 3.3.

According to Eq. (15) the optimal thickness strongly depends on the wavelength. Fig. 9 shows this in the case of biological samples, which as in this case are best represented by water. While 1 mm thick samples are very convenient at wavelengths around 1.5 Å, much thinner samples of about 20 μm thickness are required at 6 Å radiation. This roughly applies to polymer samples as well. Conducting polymers are an exception, when they contain larger amounts of heavier atoms, e.g. iodine in polyacetylene films should be some ten μm thick only for measurements at the $L_3$ absorption edge of iodine at $\lambda = 2.72$ Å. The irradiated cross section of the sample is of the order of 10 mm².

## 3.3 Estimates of Resonant Scattering

The basis for the estimation of resonant scattering and the detrimental effect of the dispersion of absorption and fluorescence on resonant dispersion terms in Eq. (29) is the optical theorem as it has been introduced by Eq. (12). All estimates given below are valid in the forward direction of scattering only. We therefore use $\sigma_v = (f'^2 + f''^2) I_v(0)$, $\sigma_{uv} = 2f'I_{uv}(0)$ and $\sigma_u = I_u(0)$.

We start with a comparison of fluorescence with resonant coherent scattering. The cross section of fluorescence $\sigma_F$ is smaller than $\sigma_{\text{photoelectric}}$ given by Eq. (12) by a factor c (c < 1) taking into account the fluorescence yield. To simplify the discussion, we assume the sample to be thin enough to show no appreciable absorption. Assuming a dilute solution of particles we can describe the resonant coherent cross section by

$$\sigma_v = 4\pi N^2 (b'^2 + b''^2) \Rightarrow 4\pi \sqrt{2} N^2 \frac{e^4}{m^2 c^4} f''^2 \tag{58}$$

This corresponds to $(f'^2 + d''^2) I_v(0)$, except that for the present pourposes we took into account the dispersionof $f'^2$ by an additional factor 1.4. N is the number of resonant atoms which form a definite structure, e.g. as metal binding sites a macromolecule. The ratio of resonant zero angle scattering to fluorescence $\sigma_F$ in forward direction is

$$\frac{\sigma_v}{\sigma_F} = \frac{4\pi \sqrt{2} N^2 b''^2}{2Nc\lambda b''} = \frac{2\pi \sqrt{2} Nb''}{c\lambda} = \frac{Nf''}{4000 c\lambda (\text{Å})} \tag{59}$$

where c is the flourescence yield shown in Fig. 10.

As b" is of the order of $10^{-10}$ mm and the wavelength of the order of $10^{-7}$ mm about some hundred atoms should scatter coherently in order give the same intensity contribution as fluorescence. This is shown as a function of the atomic number for resonance scattering at the K- and $L_3$ absorption edges in Fig. (10).

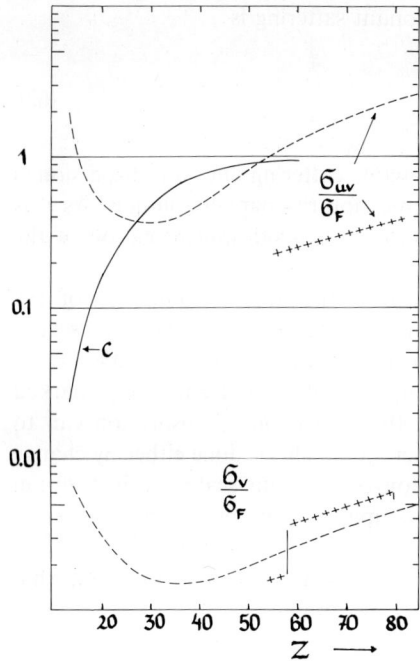

**Fig. 10.** Comparison of coherent resonance scattering with fluorescence at the — — — K-absorption edges and at the + + + $L_3$-absorption edges. The ratio $\sigma_v/\sigma_F$ is based on 20 resonant electrons (lower part). For $\sigma_{uv}/\sigma_F$ 1000 non-resonant excess electrons have been assumed. ——— c = fluorescence yield = $\sigma_F/\sigma$ absorption

To derive a similar relation for the relative contribution of the cross term f' $I_{uv}(0)$ with respect to fluorescence we introduce the contrast $\bar{\varrho}$ as the difference between the mean scattering density of the dissolved particles diminished by the scattering density of the solvent. Then the effective amplitude in forward direction of a dissolved particle is given as the product of its volume V multiplied by the contrast. With N resonant atoms per particle we obtain

$$\frac{\sigma_{uv}}{\sigma_F} = \frac{4\pi\, 2Nb'\bar{\varrho}V}{2Nc\lambda b''} \approx \frac{4\pi\bar{\varrho}V}{c\lambda} \tag{60}$$

As the variation of f' and f'' is of the same magnitude we have replaced f' by f''. This should be kept in mind whenever a more detailed consideration becomes necessary. For instance, with proteins the contrast is about 0.1 $e^-/\text{Å}^3$. Assuming a molecular volume of 10000 $\text{Å}^3$, which compares to those of small proteins with M = 10000, we obtain $\bar{\varrho}V = 2.8\, 10^{10}$ cm$^{-2}$ $10^4$ $10^{-24}$ cm$^3$ = 2.8 $10^{-10}$ cm. This corresponds to the excess scattering length of 1000 electrons. Eq. (60) then yields

$$\frac{\sigma_{uv}}{\sigma_F} \frac{4\pi \cdot 2.8 \cdot 10^{10}}{c\lambda} = \frac{0.35}{c\lambda(\text{Å})} \tag{61}$$

The above equations tell us, that the ratio between f' disperion and flourescence is proportional to the non-resonant amplitude. Fig. (10) shows that in macromolecular systems the f' term may easily exceed fluorescence.

The ratio between the f′ term and the non-resonant sattering is

$$\frac{\sigma_{uv}}{\sigma_u} = \frac{2Nb'\bar{\varrho}V}{\bar{\varrho}^2 V^2} = \frac{2Nb'}{\bar{\varrho}V} = 2\varepsilon \qquad (62)$$

This Equation tells us that the fraction of resonant scattering due to f′ dispersion is twice the ratio $\varepsilon$ between the resonant and the non-resonant amplitude. As $\varepsilon$ is much smaller than 1, the above equation is already a good approximation of the total resonant scattering.

As the fluorescence cross sections are known, the absolute intensity of the incoming beam can be determined readily from a measurement of the fluorescence intensity. On this basis all scattering data can be put on an absolute scale.

So far we were dealing with very thin samples that do not show any marked absorption. Contrary to the above relations, the dispersion of absorption can to some extent be adapted to the needs of the experiment. This is done either by change of the sample thickness or of the concentration of resonant scatterers in solution or in a solid matrix. Usually relatively thin samples will be preferred in order to decrease the absorption dispersion.

Dividing Eq. (62) by the expression given for $\mu$ in Eq. (14) and assuming that the variation of b′ can be replaced by that of b″ we obtain

$$\frac{2\varepsilon}{\mu d} = \frac{2Nb'/\bar{\varrho}V}{(\varrho N_L/M)\, 2\lambda Nb''d} \approx \frac{1}{\bar{\varrho}\lambda\varrho\, dV_{sp}} \qquad (63)$$

For a saturated hemoglobin solution at wavelengths near the K-absorption edge of iron with $\bar{\varrho} = 2 \cdot 10^{10}$ cm$^{-2}$, $\lambda = 1.74\, 10^{-8}$ cm, $v_{sp} = 0.7$ cm$^3$/g, $\varrho = 0.3$ g/cm$^3$ and $d = 0.1$ cm we obtain $2\varepsilon/\mu d = 1/7.3$, i.e. the dispersion of resonant scattering is 14% of the variation in absorption. At lower concentration, say 4% hemoglobin, the dispersion of resonant scattering and absorption are of equal magnitude. This ratio neither depends on the number of resonant scatterers bound to the particle nor on the specific scattering length b″ of the label. For dilute solutions the feasability of resonance scattering very often can be related to that of absorption edge spectroscopy, as the dispersion widths of both are comparable.

## 3.4 Requirements for Data Aquisition

To give an idea of the necessary specifations of the data aqisition system we would like to confront the reader with the following problem. What is the precision required to measure the distance between the centre of mass of a macromolecule and its resonant scattering atom at 100 Å distance? This brings us to particles with a molecular weight of about one million or more, like ribosomes. A 2% solution of the large ribosomal subunit with one resonant selenium atom would give rise to a change of the absorption at the K-edge by $\mu d = 0.02 \cdot 337.1 \cdot 0.98 \cdot 3.5 \cdot 2/1.5\, 10^6 = 3 \cdot 10^{-5}$. The dispersion of resonance scattering however is expected to be stronger, as the denominator in Eq. (63) can be decreased with respect to the situation encountered with hemoglobin in the preceding Section. The wavelength of the K-edge of selenium is only 0.98 Å, compared to 1.74 Å of the iron K-edge, the concentration is lowered to half the value, i.e. to 0.02 g/cm$^3$ and the contrast may be lowered by a factor 2 by addition of glycerol to the solvent. The overall-increase of resonant scattering then is 7.1. Resonance forward scattering of selenium in this particular situation is superior to any other kind of dispersion by a factor 7 at least. Although the relative change of the scattering due to anomalous dispersion amounts to only $7.1 \cdot 3 \cdot 10^{-5} = 2 \cdot 10^{-4}$, this experimental result would be extremely valuable as it hardly needs to be corrected for absorption, the main enemy of resonant scattering measurements.

The statistical precision of better than $10^{-4}$ can be achieved by $10^9$ scattered photons in a given interval of scattering angle. Lets divide the scattering curve into 10 intervals, then $10^{10}$ scattered photons would be sufficient to solve the above problem. Assuming that the detector would count $10^5$ useful photons per second and about the same amount of background scattering, then the 86 400 seconds of a 24 hour day at a 100% efficient synchrotron radiation source would just allow to do this experiment. Realistically one has to foresee at least two days of beam time for such an experiment.

Photon counting devices are required in order to arrive at the necessary accuracy. The noise of the detector has to be lower than 4 orders of magnitude with respect to the signal. Our experience with multiwire proportional chambers may let us believe that this rather stringent condition can be fulfilled. The simultaneous measurement of the whole scattering pattern by position sensitive area detectors is absolutely mandatory [25].

# 4 Results

So far resonance scattering in the near edge region has been measured at two laboratories: the Stanford Synchrotron Radiation Laboratory (SSRL) and the European Laboratory (EMBL) at DESY Hamburg. More recently similar experiments have been started at LURE, Orsay, France and a program on resonant scattering (including the soft X-ray spectrum) will start in HASYLAB at DESY in early 1985. In the following reference will be made to studies on non-crystalline or partially crystalline structures done at SSRL and EMBL.

## 4.1 Ferritin

This iron-storing protein has a very simple structure, which made it attractive in other pioneering work as well [31]. The spherical protein shell of about 30 Å thickness includes a mineral FeOOH core of 74 Å diameter with up to 4300 iron atoms. No wonder that the first resonant scattering experiments were carried out on this material. As the performance of the instrument X15 in june 1979 was just in the stage of many small technical improvements, the author started with a very simple problem, the dispersion of the radius of gyration R of the ferritin molecule at the iron K-absorption edge.

The radius of gyration in of dissolved particles started to play an important role, when Guinier in 1938 discovered, that small angle scattering of dilute solutions of macromolecules is described by [32]

$$I(h) = (1 - R^2h^2/3 + -)  \qquad (64)$$

and that it can most easily be determined from the negative slope of ln[I(h)] plotted versus $h^2$. As Fig. 11 shows, the dispersion of R is related to that of $f'$ [33]. There is practically no variation of R except at the absorption edge, where it increases by about 4% to a sharp peak with a full width of 11 eV at the ionization energy of one s electron of the iron atom at 7116 eV. It can be shown that the increase of R is proportional to the increase of the negative $f'$ [33]:

$$\frac{\Delta R}{R} = \frac{1}{2}\left(\frac{R_v}{R_u} - 1\right)\frac{Nb'}{\varrho V}  \qquad (65)$$

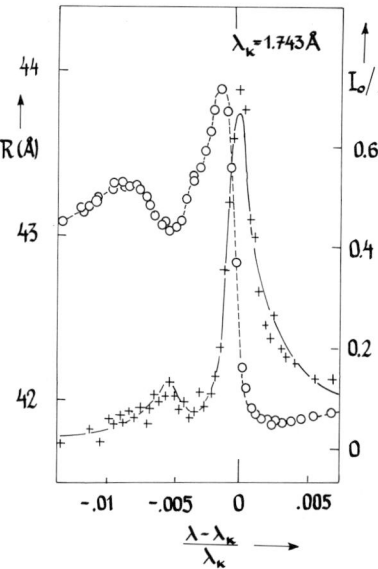

Fig. 11. The dependence of the apparent radius of gyration, R (+) of ferritin on the wavelength near the K-absorption edge. ——— as calculated from the absorption spectrum using the Kramers-Kronig relation. The absorption fine structure is clearly reflected in the dispersion of R [29]

In the off-resonance region the radius of gyration is 42 Å. This value lies well between those of iron-free apoferritin (51.5 Å) and full ferritin (28 Å) [31]. As saturated ferritin contains about 4300 iron atoms, an average iron content of about 3000 iron atom is estimated for this ferritin sample. From Eq. (65) and with reference to the radius of gyration of the FeOOH core, $R_v = 28$ Å, the relative increase of R at the K-absorption edge indicates 14% decrease of the contrast $\bar{\varrho}$ of ferritin, due to the anomalous dispersion of iron. The scattering density of the core decreases by as much as 17% and the atomic form factor of iron changes its value by one quarter (7 electrons in $f'$).

The dispersion of the absorption can readily be converted into the dispersion of $f''$ and the real resonant amplitude $f'$ can be obtained from the dispersion of $f''$ by using the Kramers-Kronig relation, Eq. (17). This serves as a test for the f' values obtained from the dispersion of R. As Fig. 11 shows there is good agreement between the experimental f' which are essentially based on the known structure of ferritin and those derived from the absorption.

Now lets turn to contrast variation by changing the scattering density of the solvent [25]. Using a 30% CsCl solution as a solvent for ferritin the real part of the complex scattering density near the $L_3$ absorption edge of caesium decreases from $0.39\ e^-/\text{Å}^3$ to about $0.37\ e^-/\text{Å}^3$ because of the negative f' of Cs, and there is also an imaginary contribution to the solvent density, which is proportional to the absorption. The excess scattering density of ferritin as a difference with respect to the solvent density then necessarly is complex as well.

$$\varrho(r) = \varrho_s(r) + (\bar{\varrho} + \bar{\varrho}' + i\bar{\varrho}'')\,v(r) \tag{66}$$

The the well-known contrast dependence of R on the real contrast assumes a more general form in the case of complex contrast [25]

$$R^2 = R_u^2 + \frac{\alpha(\bar{\varrho} + \bar{\varrho}')}{(\bar{\varrho} + \bar{\varrho}')^2 + \bar{\varrho}''^2} - \frac{\beta}{(\bar{\varrho} + \bar{\varrho}')^2 + \bar{\varrho}''^2} \tag{67}$$

where $\bar{\varrho}$ is the non-resonant contrast and $\bar{\varrho}'$, $\bar{\varrho}''$ are the resonant contributions to the contrast. The meaning of the coefficient $\alpha$ and $\beta$ is defined as usual

$$\alpha = \int \varrho_s(r)\, r^2\, d^3r/v$$
$$\beta = \int\int \varrho_s(r)\, \varrho_s(r')\, r \cdot r'\, d^3r\, d^3r'$$
$$\int \varrho_s(r)\, d^3r = 0$$
$$\int u(r)\, d^3r = v$$

These equations are a part of the more general Eq. (40). They describe the dispersion of the quadratic term of the power series of the scattering function. For spherical structures like the ferritin molecule $\beta$ is zero, as there is no change of the centre of mass at different solvent densities. Furthermore, the contributions to $\bar{\varrho}'$ and $\bar{\varrho}''$ are small compared to $\bar{\varrho}$. The radius of gyration exhibits the dispersion of f':

$$R^2 = R_u^2 + \frac{\alpha}{\bar{\varrho} + \bar{\varrho}'} \tag{68}$$

The apparent radius of gyration increases near the $L_3$-absorption edge of Cs. As the denominator $(\bar{\varrho} + \bar{\varrho}')$ in Eq. (68) increases, this observation is explained by a negative $\alpha$. From the slope of the straight line in Fig. 12 we obtain $\alpha = -0.003$. A negative $\alpha$ means that the peripheral domains of the ferritin molecule must have a lower density than the core. A similar result ($\alpha = -0.0014$) has been obtained from neutron scattering of ferritin in $H_2O/D_2O$ mixtures [31]. The difference in $\alpha$ reflects a larger intramolecular contrast in the case of X-ray scattering. This is not surprising as the scattering length of iron is so much larger than in the case of neutron scattering. At infinitely high contrast ($1/\bar{\varrho}$ in Fig. 12 tends to zero) the radius of gyration is the same for both types of contrast variation. The shape of the ferritin molecule is not changed on passing from $H_2O/D_2O$ mixtures to 30% CsCl.

Figure 12 also shows another feature, which is quite general. The operational range of contrast variation is much larger in the case of neutron scattering in $H_2O/D_2O$ mixtures. Although the difference in scattering lengths between H and D amounts to only $1.03 \cdot 10^{-12}$ cm (corresponding to 3.7 electrons difference in the atomic form factor of X-ray scattering), hydrogen is winning because of its high concentration in water. There are a 110 mol of hydrogen ion per litre compared to 2.3 mol Cs ions in the same volume. Although caesium changes its atomic form factor by 15 electrons due to the resonance scattering at the $L_3$-edge [12], its influence on the scattering density is smaller by a factor $(110/2.3)(3.7/15) = 11.8$.

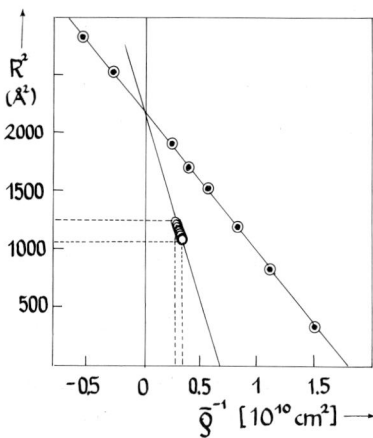

**Fig. 12.** The dependence of the apparent radius of gyration R of ferritin on the contrast. ——⊙—— Neutron scattering in $H_2O/D_2O$ mixtures. ——O—— Resonant X-ray scattering in 30%. CsCl solution near the $L_3$-absorption edge of Cs [22]

## 4.2 Parvalbumin

Contrary to the previous case, parvalbumin binds only two $Ca^{2+}$ ions. Compared to the technical difficulties encountered at the K-absorption edge of calcium due to increased absorption of 3 A wavelengths, the quantitative replacement of $Ca^{2+}$ by terbium ions ($L_3$-edge at 1.648 A) offers an ideal way to obtain structural information about the ion binding sites of parvalbumin by resonance scattering experiments. In fact, Miake-Lye, Doniach and Hodgson [34] were able to determine for the first time the distance between the center of mass of the parvalbumin and its two terbium ion binding sites in solution.

Due to details of the atomic structure involving empty d-levels [6], the lanthanides and some of the heavy transition elements have particularly strong and sharp absorption features at their $L_3$-edges ($P_{3/2}$ ionization threshold) as is shown in Fig. 13. For these elements, changes in the scattering factor of 20 electrons are observed near the peak in absorption at the edge ("white line") compared to only 3–5 electrons at most K-absorption edges and 6–10 electrons at other $L_3$-absorption edges. The fact that many of the lanthanides can replace calcium in biological systems makes them particularly attractive for the use of structural studies of calcium binding proteins.

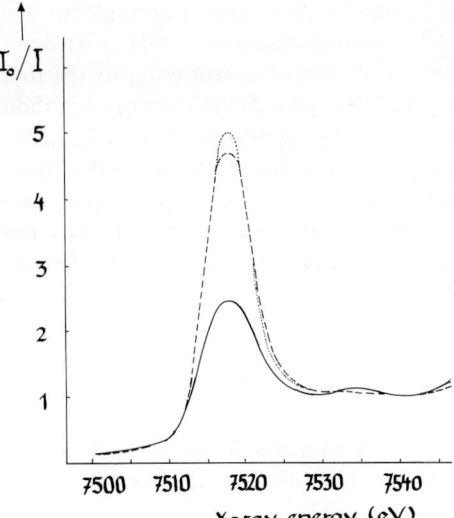

Fig. 13. X-ray absorption spectra (in arbotrary units) in a region including the terbium $L_3$ edge. The spectra were taken at SSRL beam line 1–5 using a standard transmission EXAFS set up [1], and Si 220 monochromator crystals.
........ 60 mM terbium chloride;
– – – – 60 mM terbium chloride, 20 mM EDTA;
———— terbium chloride bound to parvalbumin (30 mM).
Observe that the position of the edge does not change but the height of the absorption peak. After Miake-Lye, Doniach S., Hodgson K. O. [30]

Parvalbumin, which binds two calcium ions, was chosen for this feasability study, because it is a small protein (M = 11 500), and its refined crystal structure is known [35].

The magnitude of the changes (5%–7%) in the parvalbumin scattering curve as the X-rays are tuned through the $L_3$ absorption edge of the terbium label demonstrates that the structural information derived corresponds well with the theoretical scattering curves using coordinates derived from the crystal structure. The relative magnitude of the resonance scattering terms are simply related when they are extrapolated to zero scattering angle where all atoms scatter in phase. Based on parvalbumin's 1300 electrons in excess of the solvent level and two terbium ions the cross term $\varrho_{uv}(0)$ amounts to $2 \cdot 2 \cdot 20/1300 = 6\%$ of the total scattering. The terbium-terbium term with its 0.5% contribution to the total scattering was too for small experimental determination.

The change of the cross term $f' I_{uv}(h)$ in the small angle region is described by the dispersion of the radius of gyration given in Eq. (65). As the radial distances of the two binding sites of the terbium ions (14.4 Å and 12.7 Å from the crystallographic model) are very similar to the radius of gyration (R = 12.8 Å) difficulties were encountered to detect the small dispersion of R. The best fit of the experimental data gave $R_u$ = 12.9 Å and $R_v$ = 13.2 Å. For further work on Ca-binding proteins the SSRL reports [36] should be consulted.

## 4.3 Hemoglobin

Human hemoglobin consists of four subunits each of them having one iron atom in the oxygen binding site. What is the distance between the iron atoms in oxy-hemoglobin? As in the case of parvalbumin only the resonant crossterm $f' I_{uv}(h)$ could be measured and used for this purpose. Therefore any answer concerning iron–iron distances could be found in an indirect way only [37].

Iron bound to hemoglobin shows a slightly smaller dispersion in hemoglobin than it does in ferritin. We can count on only 4.6 resonant electrons per 2000 non-resonant excess electrons per subunit. The contribution of resonance scattering to the total scattering intensity hardly reaches 1% (Fig. 14). The change of the absorption is about 5%. This is in good agreement with the estimations given in Section 3.3. After correction for absorption and fluorescence, the scattering near the K-absorption edge follows the dispersion of f' of iron. Fig. 15 shows some dispersion curves taken at different scattering angles. The statistical accuracy of the f' dependent cross term $I_{uv}(h)$ relies on measurements of the scattering curve at thirty different wavelengths. One third of them are close enough to the resonance wavelength ($\lambda = 1.743$ Å) to show marked anomalous scattering. The statistical accuracy of $I_{uv}(h)$ as given in Fig. 14 is more or less within the dimensions of the dots. Systematic errors due to slight changes of the beam position are most serious, as they give rise to slight changes of background that cannot be corrected for.

As the concentration of 42% hemoglobin excluded any analysis at small h it was assumed that all iron atoms would have nearly the same distance from the center

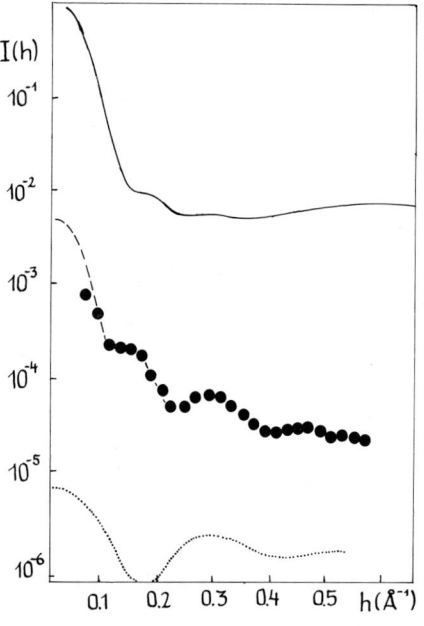 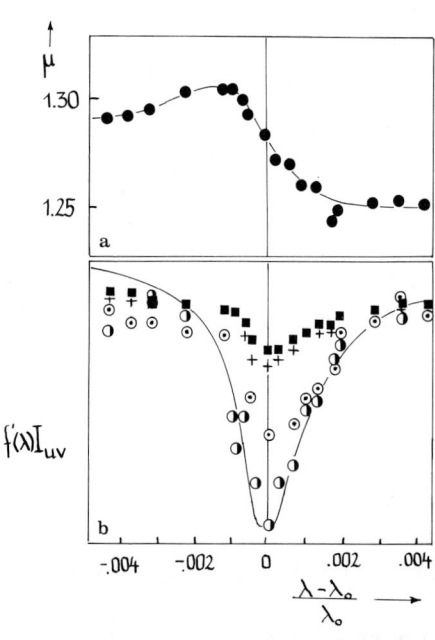

**Fig. 14.** The three basic scattering functions of oxy-hemoglobin, $I_u(h)$ (───────) f'$I_{uv}(h)$ (── ● ──), (f' + f'') $I_v(h)$ (........) as seen at the resonance peak of f' in Fig. 15. The purely resonant term due to the iron-iron distances could not be measured in this experiment but is calculated from the resulting model [33]

**Fig. 15a.** Extinction (ln ($I/I_0$)) of a 42% hemoglobin solution of 1 mm thickness near the K-absorption edge of iron at $\lambda_K = 1.743$ Å. **b** Anomalous dispersion of hemoglobin solution at various values of momentum transfer h: ◐ 0.16; ⊙ 0.18; + 0.22; ■ 0.26 respectively. ─────── f' as calculated from f'' in (a). f'' is proportional to the extinction [33]

of the hemoglobin molecule. Then the monopole scattering function has a sinusoidal shape which was fitted to $I_{uv}(h)$. From this a mean distance of 16 Å of the iron atoms from the center can be derived. After subtraction of the monopole scattering from the total scattering curve, the remainder clearly indicated a strong peak h = 0.3 which could most easily be associated with scattering from multipoles with l = 4. This leads to either planar or tetrahedral configurations of the four iron atoms. In the latter case the distance between the iron atoms of the oxyhemoglobin would be $(26 \pm 4)$ Å [37]. This is not necessarily the value which would emerge from resonance scattering studies of dilute hemoglobin solutions. In fact, the crystallographic model gives iron—iron distances around 30 Å.

## 4.4 Membranes

Apart from a few special cases where crystalline structures are formed within the membrane plane, membranes cannot be crystallized. However, membranes can be stacked in regular arrays and diffraction investigations provide structural information about the membrane. In these membrane studies, the diffraction patterns are of low to moderate resolution so that identification of particular moieties, such as amino acids, is not possible.

Reconstitution of membranes from a small number of molecular components provides simplified structures to study. Thus, cytochrome oxidase or photosynthetic reaction centers, both electron transfer proteins, may be extracted from their native membranes, purified, and reincorporated at relatively high concentration into a simple well defined lipid bilayer. Diffraction investigation then provides information about the distribution and structure of the protein in the membrane. Understanding the mechanism for electron transport in these proteins will require considerable additional information. One key element of structural informations is the location of the redox centres in the membrane profile.

Before looking at the reconstituted membranes in more detail, we shall first discuss a simpler question concerning the distribution of ions in the aqueous phase of bilayer structures. Is the ion distribution in the water phase of typically 15 Å width between the lipid surfaces homogenous, or is there a preferential binding to the polar head groups of the lipids? These studies have been started in collaboration with G. Büldt and some preliminary results from dipalmitoyl phosphatidyl cholin (DPPC) membranes will be reported here [38].

To trace the ion distribution across the membrane profile, erbium ions were chosen, because of the strong anomalous dispersion at the $L_3$-absorption edge which they share with other lanthanide ions. The concentric rings of diffraction were recorded with a position sensitive area counter up to the 9th order. The 0.5 molar concentration of erbium ions in the water phase (20% of the sample volume) gave rise to typically 5 to 10% change of the diffraction due to resonance scattering in the $L_3$-absorption edge (Fig. 16). The magnitude of dispersion of absorption and fluorescence were nearly equal: scattering decreased by a factor 2 in the absorption peak of the $L_3$-edge and fluorescence compensated for this loss at large scattering angles.

Similar experiments have been done at a lower (0.18 M) and higher (1.8 M) concentration of erbium ions.

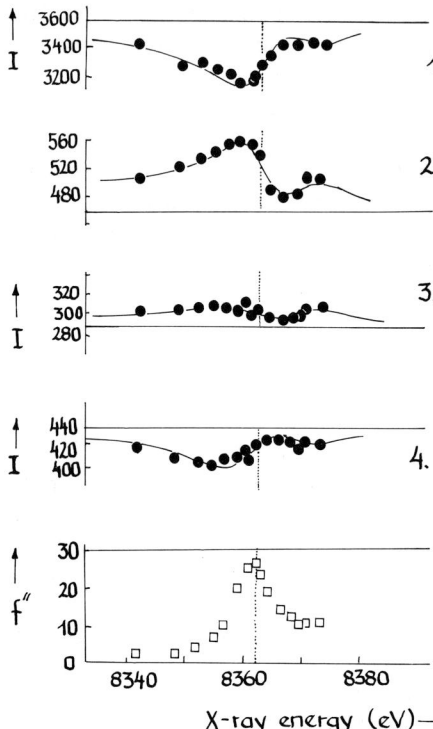

**Fig. 16.** Resonance scattering of erbium ions incorporated in the water phase (20%) of dipalmitoyl diphospatidyl cholin (DPPC) near the $L_3$ absorption edge of erbium at $\lambda = 1.482$ Å) [34]. The intenity units are proportional to the number of photons per ring of the Debye-Scherrer diagramme

The analysis of the dispersion of the peak intensities starts with Eq. (29). As the periodic structure extends only across the membranes, this equation reduces to

$$S(h) = A^2(h) + 2 f' A(h) B(h) + (f'^2 + f''^2) B^2(h) \qquad (69)$$

where $h = n\, 2\pi/d$, $n = 1, 2, 3, \ldots$ and d is the spacing of the unit cell. The knowledge of the sign of the cross term is useful for the determination of the sign of $S_u(h)$, provided we can make some reasonable guess on the sign of the resonant amplitude B(h). In fact, the labelled region is restricted to the water phase, which is only 1/5 of the repeat unit. This supposedly well-defined structure in physical space leads to a more reliable prediction of the signs of B(h) than this is the case for the non-resonant amplitude A(h). From contrast variation of the water phase with neutron scattering [39] it is known that B(h) is positive in its first four orders. Resonant scattering of erbium ions seems to indicate a change of the sign of B on passing from the third to the fourth diffraction peak. A consequence of this finding is that the ion distribution is different from the water distribution. A tendency of the erbium ions to bind to the polar head groups of the lipids is obvious [38].

Now lets return to reconstituted membranes. Many redox centers contain metal atoms. If these could be isomorphically replaced by atoms of substantially different scattering power, then measurement of changes in diffracted lamellar intensities even at low resolution combined with the prior knowledge of the number of metal binding

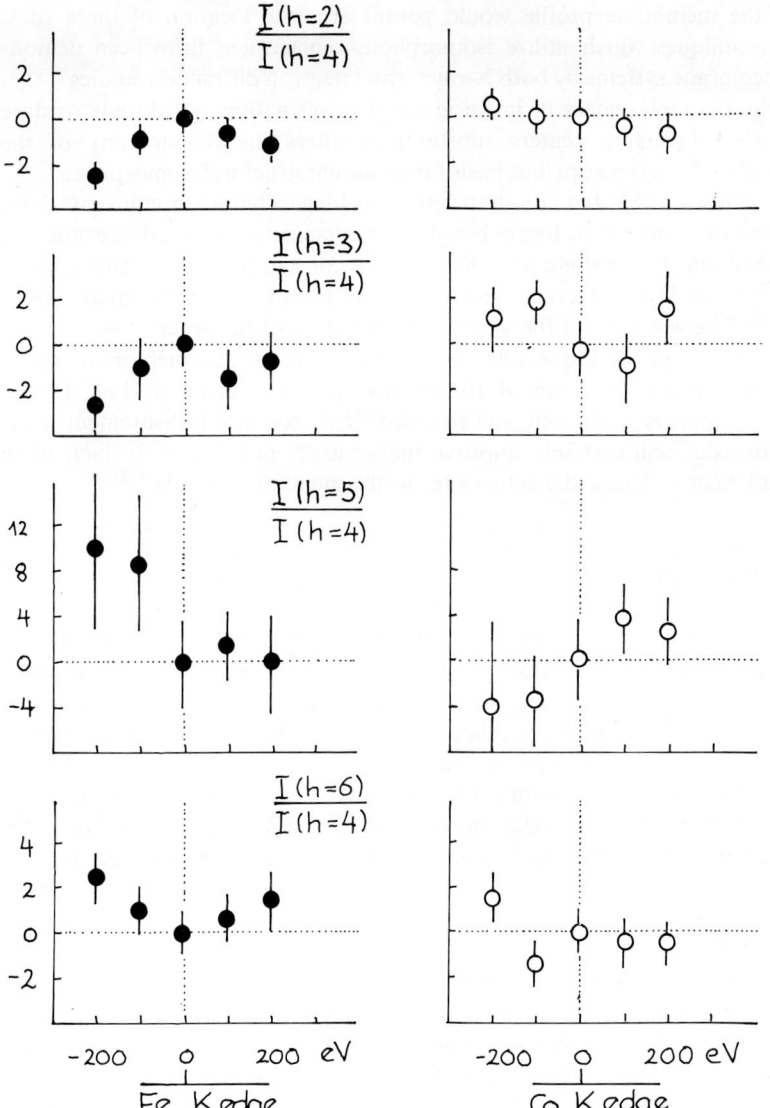

**Fig. 17.** The ratios of the integrated intensities of the lammelar reflections $R(h/h') = ([S(h)/S(h')]_1 - [S(h)/S(h')]_2)/[S(h)/S(h')]_2$ for a reconstituted cytochrome oxidase membrane multilayer. The subscript 1 denotes the incident X-ray energy and the index 2 corresponds to an energy far from the absorption edge. $h' = 4$, $h = 2, 3, 5$ and 6 at incident X-ray energies near the iron K-absorption edge (●). The f' dispersion is most significant with $h = 2$ and $h = 6$. There is no dispersion at wavelenghts near the Co K-absorption edge (○). After Stamatoff et al. [39]

sites within the membrane profile would permit accurate location of these sites. Diffraction techniques which utilize isomorphous replacement have been demonstrated for membrane systems by both X-ray [40] and neutron diffraction studies [41,42]. Unfortunately, the replacement of intrinsic metal atoms within cytochrome oxidase or photosynthetic reaction centers substantially alters the functionality of the membrane so that there is no obvious basis for assuming structural isomorphism.

X-ray resonant scattering as a non-destructive labelling technique circumvents these problems. First experiments to locate metal atoms associated with redox centers in biological membranes were done at SSRL [43]. The concentration of intrinsic metal atoms in these membranes formed with cytochrome oxidase is very small (about 1 atom in $10^4$). The variation of the scattering intensity is of the order of $10\%$ at the iron K absorption edge for the reconstituted cytochrome oxidase membrane multilayer. Even at the wide spacing of the energy intervals shown in Fig. 17 the dispersion of $f'$ appears to be well well resolved [43]. Resonance measurements near the absorption edge will certainly improve the accuracy in $f' S_{uv}(h)$. In fact, these data were sufficient to locate the active sites in the membrane profile [44].

## 4.5 Polyelectrolytes

Depending on the density of charges along the polymer chain, these macromolecules may display a wide variety of conformations. Ionomers for instance are polyelectrolytes which carry only few, randomly distributed charges. Their use as thermoplasts is due to their properties which are related to those of uncharged polymers and of highly charged polyelectrolytes as well. There are polyanions, polycations and polyampholytes. Well-known polyanions are polyvinyl and polystyrolsulfonic acid, polyacrylic and polymethacrylic acid, polyphosphoric acids etc. Examples of polycations are polyvinylamin and poly(4-vinylpyridin) in acid solution and poly(4-vinyl-N-dodecylpyridinum) salts. Polyampholytes are usually copolymers, e.g. methacrylic acidvinylpyridin, showing both basic and acid properties depending on the solvent. Semisynthetic products, mostly derivatives of polysaccharides, like carboxymethylcellulose are of continuing interest.

The greatest variety of polyelectrolytes is found in nature. The role of electric charges is essential for the proper functioning of nucleic acids, the numerous enzymes, proteins and polysacchardies. The fundamental role of polyelectrolytes in all living processes is unquestionable.

Polyelectrolytes in many ways show an unusual behavior, which is due to high charge density concentrated on the chain. There is a strong interaction between the segments of the same chain and other macromolecules which clearly indicates a long range electrostatic potential. The latter has its image in the counterion distribution around charged polymer chains. The conformational and folding properties of polyelectrolytes and their specific functioning could be better understood once the spatial distribution of the various kinds of free ions of the solvent could be determined.

Methods of isomorphous replacement usually fail to give any reliable structural information as the exchange of counterions may modify the structural appearance of the charged macromolecule considerably. Isotopic replacement of hydrogen by deuterium in tetramethylammonium counter ions has proven to maintain isomorphism within the accuracy of low resolution structural studies using neutron small angle scattering [45]. Resonance scattering in fact provides a unique way of non-destructive labelling each kind of ions separately just by tuning the wavelength to the absorption edge of the ion in question.

The first experiments on resonant scattering of counterion distributions fall into the early period of the operation of the EMBL instrument X15. Among the studies on polyacrylic acid, desoxyribonucleic acid (DNA) and ribosomes in various solvents, those on Cs-DNA gave the clearest answer with respect to the counter ion distribution. Another advantage was that the results could be compared with the existing double helic model with its predictable counter ion distribution [46].

As the DNA double helix forms a rather stiff rod-like structure, the spatial distribution of the counterions is described by a radial function. The dispersion of the radius of gyration of the cross section $R_q$ [47] reflects the distribution of the caesium ions as one measures the scattering profile at various wavelengths near the "white line" of the caesium $L_3$ absorption edge, which is shown in Fig. 18 [17]. There is a change of the radius of gyration of the cross section from 9.8 Å to 9.3 Å due to the dispersion of f' of the caesium ions. The decrease of the radius of gyration near the absorption edge tells us that according to Eq. (65) the radial distribution of the caesium ions must have a greater radius of gyration $R_q$ of the DNA molecule in $H_2O$ amounts to only 7.2 Å [48]. If the relative change of the scattering intensity at small angles due to resonance scattering had been determined with reasonable accuracy, then from the ratio between resonant and nonresonant excess electrons per unit length of the double helix the rms distance of the caesium ions from the axis of the DNA double helix could have been determined by using Eq. (65). Caesium ions approach the double helix axis up to 12 Å [46,49].

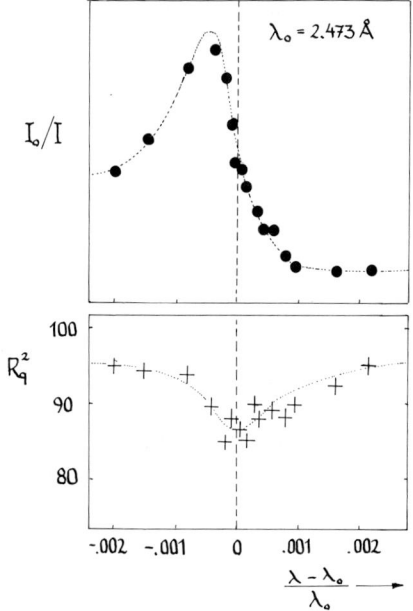

**Fig. 18.** The anomalous dispersion of the cross section radius of gyration of Cs-DNA double helix. The depression of $R_q$ is due to a decrease of the scattering factor of Cs at the $L_3$-absoption edge. (●) absorption spectrum of Cs, (+) dispersion of $R_q$ of Cs-DNA [42]

## 4.6 Conducting Polymers

For a polymer to be conducting, some electrons in the backbone must be less strongly localized in the chemical bonds. This is possible if the polymer has an unsaturated or conjugated structure.

The simplest polymer with a conjugated backbone is polyacetylene. Its structure is similar to that of the saturated polymer polyethylene, but has one of the hydrogen atoms removed from each carbon of the polyethylene chain. Each carbon atom in the polyacetylene chain thus has one excess electron which is not involved in the basic chemical binding. And if the separation of the carbon were constant, polyacetylene would conduct along the chain; in other words it would behave like a metal in one dimension. But unfortunately this is not true as the "free" electrons tend to get localized in shorter double bonds. Conjugated polymers can at best be expected to display semiconducting properties.

Quite small quantities of dopant — additional electron accepting molecules, such as the halogens, bromine, chlorine, iodine, or arsenic pentafluoride — causes the conductivity of trans-polyacetylene to increase from $10^{-5}$/ohm cm to over 500/ohm cm, the final value being close to that of metals. The microscopic processes for the properties are still a matter of debate, as the films have a complex morphology. Two alternative views are discussed: the model of a one-dimensional conductor, in which molecular dopants act like dopants in a conventional semiconductor, producing either an excess or a deficiency of electrons in the backbone. The alternative model rejects a purely one-dimensional view as it could be shown that the initial doping of polyacetylene is inhomogeneous, giving local regions of high conductivity and that the motion of the "excess electron" is not one-dimensional [50].

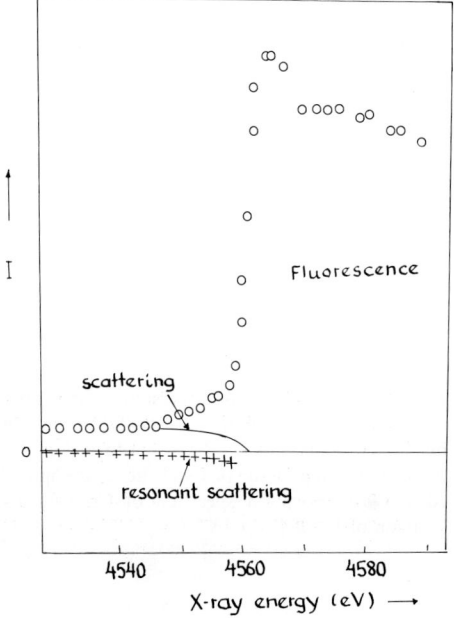

Fig. 19. Resonant scattering, absorption and fluorescence of iodine in a 100 μm foil of polyacetylene, $CHJ_{0.1}$, near the $L_3$-absorption edge ($\lambda = 2.72$ Å) at h = 0.1 $A^{-1}$. The very high dispersion of fluorescence and absorption at the short wavelength side of the absorption edge limits the measurement of resonance scattering to the pre-edge region of the spectrum. (+) f'(E) $S_{uv}$

At this point first preliminary resonance scattering experiments on polyacetylene foils doped with iodine to nearly saturation have been performed at wavelengths near the $L_3$ absorption edge of iodine ($\lambda = 2.72$ Å) in collaboration with C. Riekel. The scattering pattern from a 100 μm thick foil resulted in a nearly monotonous decrease of the intensity at h values between 0.1 $A^{-1}$ to 1.0 $A^{-1}$ (60 Å to 6 Å resolution). The resonant scattering of iodine as it appears in $f'I_{uv}(h)$ is compared with the much stronger variation of absorption and fluorescence shown in Fig. 19. Unfortunately the more structured intensity distribution at higher h-values cannot be reached with the existing instrument X15 at EMBL. The new instrument presently under construction at HASYLAB will achieve a resolution which is equal to the wavelength actually used. This will be possible with a system of three area detectors in an asymmetric arrangement, which covers scattering angles up to 60°. The combined use of EXAFS (see Sect. 2.2) and anomalous scattering will cover pair correlations from nearest neighbours to several hundred Å. The wide choice of elements, the structure of which can be looked at separately, makes these two techniques to most promising tools in structural studies of conducting polymers.

## 5 Conclusions

Resonance scattering is both a method of molecular structure determination and a chemical analytical tool as well. It identifies and selects specifically the coherent scattering of any chemical element. Resonance scattering distinguishes clearly between the intensity due to the structure of the excited atoms only and those contributions

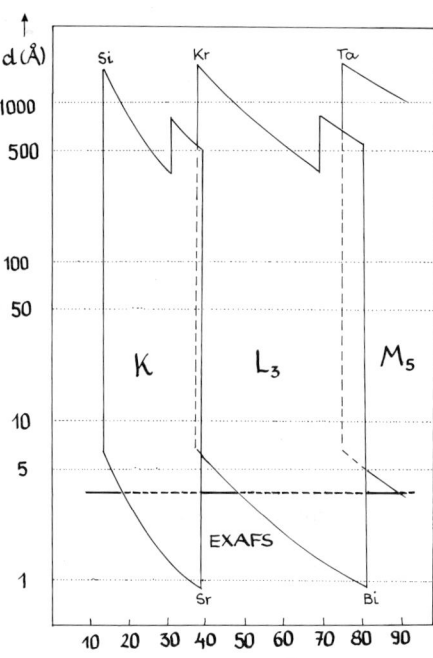

**Fig. 20.** The domains in which X-ray resonance scattering and EXAFS provide structural information. Structural resolution from diffraction is assumed to be limited by the maximum scattering angle of 60° and the wavelength range from 0.8 to 7.4 Å. This compares to the specifications of the instrument given in Fig. 8

which are due to the convolution of the former with the structure of the nonresonant atoms. This distinction is less clearly obtained from EXAFS measurements [20]. The partial structure of resonant atoms as the structure of interest is obtained cleanly by resonant scattering. It may be used either as a very sensitive and specific probe of macromolecular rearrangements or as a basis for the elucidation of the total structure. The latter step is accomplished by using the distinct convolution term of resonant scattering in addition. This strategy works succesfully with single crystal structure analysis. Its adaptation to the study of disordered or partially ordered structures is only at the beginning.

Considerable progress is expected once the instrument developement allows the full exploitations of the spectral brilliance of synchrotron radiation. Major improvements will have to come from the data aquisition system. The state of art of selecting monochromatic radiation out of the continuoss spectrum of synchrotron radiation is relatively better, though the possibility for the realization of more efficient systems still exists. Assuming that the technical developpements envisaged in various synchrotron radiation laboratories will take place, the above strategy of data analysis (Sect. 2) can come to work. The domains in which X-ray resonance scattering and X-ray spectroscopy provide structural information are given in Fig. 20.

*Acknowledgements*: The Resonant X-ray Scattering Project started at the Outstation of the European Molecular Biology Laboratory at DESY Hamburg. Many ideas which have been expressed in this article have come up with the plan to include soft X-ray resonant scattering in macromolecular structure research, a project of the University of Mainz at the Hamburger Synchrotronstrahlungslabor HASYLAB at DESY which is supported by the Bundesminister für Forschung und Technologie under the grant application nr. 05 294SN. Writing this text was greatly facilitated by a word processor from the Verband der Chemischen Industrie. To all of them I would like to express my sincere thanks.

# 6 References

1. Ramaseshan, S., Ramesh, T. G., Ranganath, G. S.: Anomalous Scattering, Ramaseshan, S., Abrahams S. C. (ed.), Munksgaard, Copenhagen, 1974
2. International Tables of Crystallography, 3, 157, The Kynoch Press, Birmingham, 1968
3. Schäfer G. F., Fischer K.: Z. F. Kristallographie, 62, 273 (1983)
4. Lye, C., Phillips, J. C., Kaplan, D., Doniach, S., Hodgson, K. O.: Proc. Nat. Acad. Sci. U.S.A. 77, 5884 (1980)
5. Templeton, D. H., Templeton, L. K., Phillips, J. C., Hodgson, K. O.: Acta Cryst. Sect. A 36, 436, (1980)
6. Guinier A.: Les Rayons X, Presses Universitaires de France 1984
7. Winick, H., Doniach, S., (ed.): Synchrotron Radiation Research, Plenum Press, New York, (1980)
8. Bordas, J.: Uses of Synchrotron Radiation in Biology, Stuhrmann, H. B. (ed.) Academic Press, London, (1982)
9. Woolfson M. M.: An Introduction to X-Ray Crystallography, Cambridge University Press, (1970)
10. Okaya, Y., Saito, Y., Pepinsky, R.: Phys. Rev., 98, 1857 (1955)
11. Bragg W. L., Perutz, M. F.: Acta Cryst. 5, 277, (1952)
12. Blow, D. M., Crick, F. H. C.: Acta Crystallogr. 12, 794, (1959)

13. Phillips, J. C., Hodgson, K. O.: Synchrotron Radiation Research, Winick H., Doniach S. (ed.), Plenum Press, New York, (1980)
14. Bartunik, H., Fourme R., Phillips, J. C.: Uses of Synchrotron Radiation in Biology, Stuhrmann, H. B. (ed.), Academic Press, London (1982)
15. Glover, I. D., Tickle, I. J., Wood, S. P., Pitts, J. E., Blundell T. L., Bartels K. S., Bartunik, H.: in press
16. Fischer, K.: Z. Naturforsch. 36a, 1253 (1981)
17. Stuhrmann, H. B.: Quarterly Review of Biophysics, 14, 433 (1981)
18. Hendrickson, W. A., Teeter, M.: Nature 290, 107 (1981)
19. Stuhrmann, H. B.: Acta Crystallogr., Sect. A 26, 297 (1970)
20. Svergun, D. I., Feigin, L. A., Schedrin B. M.: Acta Crystallogr., Section A, 38, 827 (1982)
21. Stuhrmann, H. B.: Z. Phys. Chem. Frankfurt, 72, 177 (1970)
22. May, R. P., Stuhrmann, H. B., Nierhaus, K. H.: Neutrons in Biology, Schoenborn, B. P. (ed.), Plenum Press, 1984
23. Fuoss, P. H.: SSRL Report 80/06, Stanford Synchrotron Radiation Laboratory, Standford, California, (1980)
24. Keating, D. T.: J. Appl. Phys., 34, 923 (1963)
25. Stuhrmann, H. B., Gabriel, A.: J. Appl. Cryst., 16, 563 (1983)
26. Zachariasen, W. H.: Theory of X-Ray Diffraction in Crystals, Dover Publications, New York
27. Beaumont W., Hart, M. J.: Appl. Phys. E7, 823 (1974)
28. Howell, J. A., Horowitz, P.: Nucl. Instr. Meth. 125, 225 (1975)
29. Stuhrmann, H. B.: Die Makromolekulare Chemie, 183, 2501 (1982)
30. Materlik, G.: personal communication
31. Stuhrmann, H. B., Haas, J., Ibel, K., Koch, M. H. J., Crichton, R. R.: J. Mol. Biol. 100, 399 (1976)
32. Guinier A.: Ann. Phys. 12, 161 (1939)
33. Stuhrmann H. B.: Acta Crystallogr., Sect. A 36, 996 (1980)
34. Miake-Lye, R. C., Doniach, S., Hodgson, K. O.: Biophysical J., 41, 287 (1983)
35. Kretsinger, R. H., Nickolds, C. E.: J. Biol. Chem., 248, 3313 (1973)
36. Hubbard, S. R., Miake-Lye, R. C., Doniach, S., Fairclough, R. H., Hodgson, K. O.: SSRL Report 1984, proposal No 712B, IX-57, Stanford (1984)
37. Stuhrmann, H. B., Notbohm, H.: Proc. Nat. Acad. Sci. U.S.A., 78, 6216 (1981)
38. Stuhrmann, H. B., Büldt, G.: manuscript in preparation
39. Worcester, D. L.: Biol. Membranes, 3, 1 (1976)
40. Stamatoff, J., Bilash, T., Ching, Y., Eisenberger, P.: Biophys. J., 28, 413 (1979)
41. Blasie, J. K., Schoenborn, B. P., Zaccai, G.: Brookhaven Symposium n Biology, vol. 27
42. King, G. I., Mowery, P. C., Stoeckenius, W., Crespi, H. L., Schoenborn, B. P.: Proc. Nat. Acad. Sci. U.S.A., 77, 4726 (1980)
43. Stamatoff, J., Eisenberger, P., Blasie, J. K., Pachence, J. M., Tavormina, A., Erecinska, M., Dutton, P. L., Brown, G.: Biochim. Biophys. Acta, 679, 177 (1982)
44. Blasie, J. K., Pachence, J. M., Tavormina, A., Erecinska, M., Dutton, P. L., Stamatoff, J., Eisenberger, P., Brown, G.: Biochim. Biophys. Acta 679, 188 (1982)
45. Ragnetti, M.: Thesis, University of Mainz, 1984
46. Oberthür, R. C.: Thesis, University of Mainz, 1974
47. Kratky, O., Pilz, I.: Quarterly Rev. Biophys., 5, 481 (1972)
48. Luzzati, V, Nicolaieff, A., Masson F.: J. Mol. Biol. 3, 185 (1961)
49. Clementi, E., Corongiu, G.: J. Biol. Physics, 11, 33 (1983)
50. Wegner, G.: Angew. Chemie, 93, 352 (1981)

H.-G. Zachmann (Editor)
Received June 6, 1984

# X-ray Line Shape Analysis.
# A Means for the Characterization of Crystalline Polymers

Géza Bodor
Polymer Research Institute, H-1950 Budapest, Hungary

*The profile analysis of WAXS data yields informations in term of crystallite size, microstrain and about the crystallite size and strain distribution.*
  *The average crystalline particle size is derived from the half width line broadening, all other results are calculated from the parameters of the investigated line profile.*
  *The line broadening of the experimental profile consists always of an instrumental line broadening and a structural line broadening. The former could be measured separately by using standard samples, which have infinite large crystallites and have no strains.*

1 Structural Broadening . . . . . . . . . . . . . . . . . . . 166

2 Calculations from the Structural Broadening . . . . . . . . . . . . . . . 167

3 Calculations from the Distortion and from the Particle Size Fourier Coefficients 186

4 Main Symbols Used in the Text . . . . . . . . . . . . . . . . . 191

5 Summary . . . . . . . . . . . . . . . . . . . . 192

6 References . . . . . . . . . . . . . . . . . . . . 192

## 1 Structural Broadening

The structural broadening is the subject of our calculations. First we want to show, how we can get the structural broadened line profile from the experimental data.

The h(x) experimental profile is a convolution of the g(x) instrumental and of the f(x) structural profile:

$$h(x) = f(x) * g(x) \quad \text{or} \tag{1/a}$$

$$h(x) = \int_{-\infty}^{\infty} f(y)\, g(x-y)\, dy \tag{1/b}$$

where h(x) is the intensity measured at distance x. The value of y is a distance in the direction x and used for the calculations.

A deconvolution process is necessary to calculate the line profile, caused by structural broadening. Such a deconvolution process was proposed by Stokes [1] which calculates the H(t) and G(t) Fourier coefficients of the h(x) experimental and of the g(x) instrumental broadenend line profile and from these the F(t) Fourier coefficients of the f(x) structural broadened profile.

It should be emphasized, that the Stokes method don't assume any shape or equation of the experimental data.

The H, G and F are the Fourier transforms of the h(x), g(x) and f(x) functions, whose values are given by Equations

$$H(t) = \frac{1}{a} \int_{-a/2}^{a/2} h(x) \exp(2\pi\, ixt/a)\, dx \tag{2}$$

$$G(t) = \frac{1}{a} \int_{-a/2}^{-a/2} g(x) \exp(2\pi\, ixt/a)\, dx \tag{3}$$

$$F(t) = H(t)/a\, G(t) \tag{4}$$

where t is the harmonic number of the Fourier coefficients, a is the intervall of the calculations.

The f(x) structural broadened line profile could be calculated

$$f(x) = \sum \frac{H(t)}{aG(t)} \exp\{(-2\pi\, ixt)/a\} \tag{5}$$

(Only the cosine transforms are to be considered, if the functions are symmetrical about their origin points.)

The representation of the F(t) Fourier coefficients of the structural broadened peak shape by a plot of F(t) vs. t gives a good method for representing different effects in the sample.

To illustrate this, a plot of the coefficients, originally presented by Warren and Averbach [2] is shown in Fig. 1.

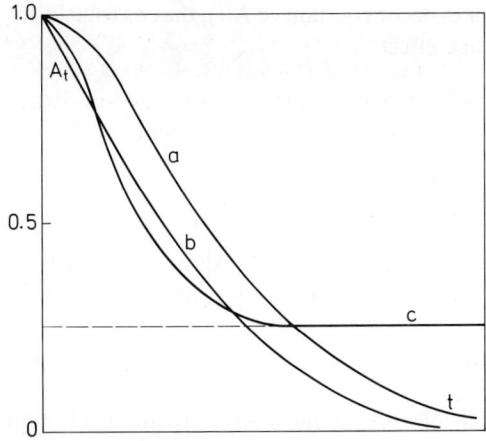

**Fig. 1.** Difference in the shape of the $A_t(n)$ vs. t curves. *a)* Cold work distortion *b)* particle size broadening *c)* temperature effect. (Warren and Averbach [2])

The cold work distortion is noted *a*, the particle size broadening *b*, the temperature effect *c*. Any kind of wrong data disturbs the smoothness of the coefficients.

## 2 Calculations from the Structural Broadening

It is generally accepted, that the structural broadening consist of distortion broadening and of particle size broadening. The main question is, what kind of information from the structural broadening we are able to get.

The integral breadth of the lattice distortion profile, according to Wilson [3]:

$$\beta_D = \frac{4e \sin \theta}{\lambda} = \frac{2e}{d_{hkl}} \tag{6}$$

where $e = \Delta_{hkl}/d_{hkl}$ is a measure of the maximum lattice distortion.

The integral breadth of the crystalline size profile is taken as the Scherrer Equation [4]:

$$\beta_S = \frac{K}{D_{hkl}} \tag{7}$$

where K is a constant, related both to crystallite shape and to the way in which $\beta_S$ and $D_{hkl}$ are defined. We take $\beta_S$ in (Å$^{-1}$), $D_{hkl}$ (in Å) and K to be an integral breadth, the dimension of the crystal normal to the (hkl) plane, and unity, respectively.

The Equations (6) and (7) are most commonly applied in an alternative form:

$$\beta_D(\text{rad}) = \beta_D(\text{Å}^{-1}) \lambda/\cos \theta = 4 e \tan \theta \tag{8}$$

$$\beta_S(\text{rad}) = \beta_S(\text{Å}^{-1}) \lambda/\cos \theta = \frac{K \cdot \lambda}{D_{hkl} \cos \theta} \tag{9}$$

The F(t) Fourier coefficients of the Stokes deconvolution or A(t), the cos transforms are the products of the coefficients of each effect:

$$A_t(n) = A_t^S \cdot A_t^D(n) \tag{10}$$

where A is the cosine part of the Fourier coefficient,
 t is the harmonic number of the coefficient and
 n is the order of the reflexion.

The $A_t^S$ particle size coefficient is independent of n, while $A_t^D(n)$, the distortion coefficient is n dependent. A plot of ln $A_t(n)$ vs. $n^2$ gives a slope and an intercept:

$$\ln A_t(n) = \ln A_t^S + \ln A_t^D(n) \tag{11}$$

This is the so called Warren-Averbach plot. $A_t^S$ is the portion of the coefficient due to mosaic size.

In the case of a strain, in each direction, for each t there is a $\delta_t$ displacement. This $\delta_t$ displacement is usually expressed in terms of its components along different directions. Perpendicular to direction d, the relative displacement is $Z_t = \delta_t/d$. The average over all relative displacements, for all harmonic numbers in the direction perpendicular to d is $\langle Z_t \rangle$, the reduced displacement.

It has been shown by Warren and Averbach, that the cos distortion coefficient is:

$$\ln A_t^D(n) = -2\pi^2 n^2 \langle Z_t^2 \rangle \tag{12/a}$$

As the harmonic number t corresponds to a separation $R = t \cdot d$, this equation is sometimes written in the form

$$\ln A_t^D(n) = -2\pi^2 n^2 t^2 \langle \varepsilon_R^2 \rangle \tag{12/b}$$

or

$$\ln A_t^D(n) = -2\pi^2 n^2 \langle \varepsilon_R^2 \rangle R^2/d_{n=1}^2 \tag{12/c}$$

where the term $\langle \varepsilon_R^2 \rangle$ is the mean square microstrain averaged over all distances R, normal to the diffracting planes, d is the interplanar spacing for the first order peak.

The reduced displacement $\langle Z_t^2 \rangle$ and the relative mean square microstrain $\langle \varepsilon_R^2 \rangle$ are connected by the equation

$$\langle Z_t^2 \rangle = t^2 \langle \varepsilon_R^2 \rangle = \frac{R^2}{d_{n=1}^2} \langle \varepsilon_R^2 \rangle \tag{13}$$

There are errors in the first few $A_t(n)$ values for any peak due to difficulties in establishing the correct background for each peak. This leads to a "hook" in $A_t^S$ versus t or R near t = 0, which can corrected by plotting ln $A_t(n)$ versus t or R before the use of Eq. (11). This should be linear over an appreciable range of t or R. (See the remarks about the hook effect made by Delhez and coworkers, [22] explained later on.)

From this point, different interpretations are in the literature. In order to try to give a summary, the Table 1. shows the probably most important publications on the field.

First the integral breadth methods, then the Fourier transform methods and then the crystalline particle size distribution methods are listed, on a time scale, taken in account, how many orders of reflexion are needed for the evaluation.

The first Fourier coefficients of a polymer diffraction pattern were determined by Katayama [5]. His work was reinterpreted in two papers and we shall return to his data later on.

The theory of the paracrystal was used to develop a method to distinguish particle size and distortion, proposed by Bonart, Hosemann and McCullogh [6]. By measuring the line broadening of at least two orders of reflexion, the value of D, the crystalline particle size and g, the paracrystalline distortion fluctuation parameter could be adjusted so, that g becomes a constant for the observed reflexions.

Thielke and Billmeyer were assuming, that broadening due to small crystal size and that due to imperfections are additive [7]. The corrected line breadth is given by

$$\frac{B \cos \theta}{\lambda} = \frac{1}{D} + \frac{\mu \sin \theta}{\lambda} \tag{14}$$

where D is the apparent crystal size and $\mu$ is the breadth of a strain distribution characterizing chain imperfections. Using at least two orders of reflexions, the equation could be solved. For PE single crystal mats they observed for the two hk0 planes 200 Å crystallite size and 1.3% strain. Assuming the same strain for the 002 plane, the crystallite size was calculated to be 105 Å, in good aggrement with the lamellar thickness observed by low angle X-ray diffraction. They concluded, that firm conclusions cannot be drawn about the origins of line broadening in polymer crystals unless two, or preferably more successive orders of reflexion can be observed. In the absence of multiple orders, it is virtually fruitless to attempt to assign the origins of line broadening to strain or imperfections, small crystallite size or a combination of these factors.

A breadth analysis was proposed by Schoening to get the strain and particle size values from X-ray line breadths [8]. Assuming, that the particle size broadening tends to give a Cauchy line shape and strain broadening a Gauss shape, the observed intensity profile is a convolution of a Cauchy and a Gauss profile. The integral breadth contains of the line breadths of both profile and from two orders of reflexions from the same plane these could be determined. As the formulas are rather complicated and long, Schoening gives a figure and a table for the determinations.

A detailed summary was presented by C. N. J. Wagner about the analysis of the broadening and changes in position of peaks in X-ray powder pattern [9].

The first separation of size and distortion effects on the X-ray pattern of a polymeric system was completed by Buchanan and Miller [10].

The methods, proposed till that time (1966) were applied to isotactic polystyrene, which shows sharp diffraction maxima, including up to six orders of (110) reflexion.

Fourier transform methods, according to Warren and Averbach [2] were tryed. The extrapolation results are presented in Fig. 2. The upward concavity of the curves implies, that the strain distribution falls off more slowly, then does a Gaussian.

**Table 1.** Chronological order of line shape analysis

| Integral breadth methods | | | Fourier transform methods | | | Crystalline particle size distribution | |
|---|---|---|---|---|---|---|---|
| Reflexions of 3 Order | 2 Order | 1 Order | Reflexions of 3 Order | 2 Order | 1 Order | Reflexions of 2 Order | 1 Order |
| | Bonart, Hosemann, McCullogh 1963 Thielke, Billmeyer 1964 | | | Warren-Averbach 1952 Katayama 1961 | | Bertaut 1952 | Doi 1961 |
| | Schoenig 1965 | | | | | | |
| Buchanan Miller 1966 | | | Buchanan Miller 1966 Harrison 1967 | Wagner 1965 | | | |
| | | | | Takahashi 1969 Kulshresta Dweltz 1971 | | | |
| | | | | | | | Yoda, Doi Tamura Kuriyama 1973 |
| Vogel, Haase, Hosemann 1974 | | | Wecker Cohen Davidson 1974 | Vogel, Haase Hosemann 1974 | Gangulee 1974 | | |

|  |  |  |  | Yoda, Tamura Doi 1976 Yoda, Kuriyama 1977 |
|---|---|---|---|---|
|  |  |  | Bodor Füzes 1978 |  |
|  |  |  |  | Odijama, Noto Yamane, Ishibashi 1980 |
|  |  |  | Bodor Füzes 1982 |  |
|  |  | Mignot, Rondot 1975 |  |  |
|  | Crist, Cohen 1979 |  |  |  |
|  |  | Delhez De Keijser 1982 |  |  |
| Langford 1978 |  |  |  |  |
| W. Schmidt 1980 Krenzer Ruland 1981 |  |  |  |  |
| De Keijser, Langford 1982 |  |  |  |  |

From the intercepts on the ordinate, $A_t^s$ values were calculated. These particle size coefficients (gained from 3 iPS samples) are plotted in Fig. 3 against $R = a_3 t$. The intercepts on the abscissa gives $\bar{D}_{hkl}$, the average crystalline particle size. In each case, renormalisation was necessary, indicating the difficulty of measuring the correct profile tails.

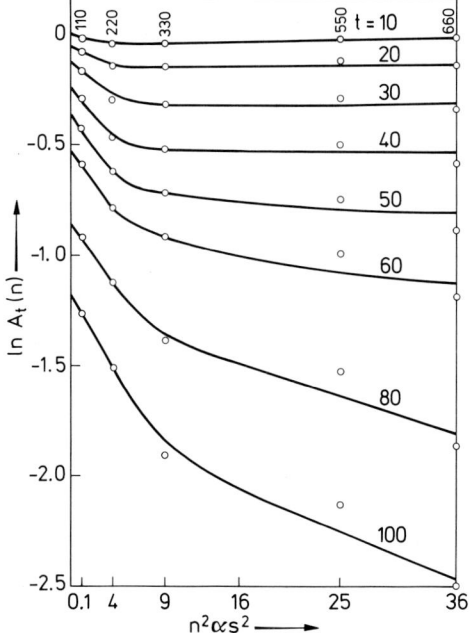

Fig. 2. Extrapolation curves for the Warren-Averbach Fourier transform method. (Buchanan-Miller [10])

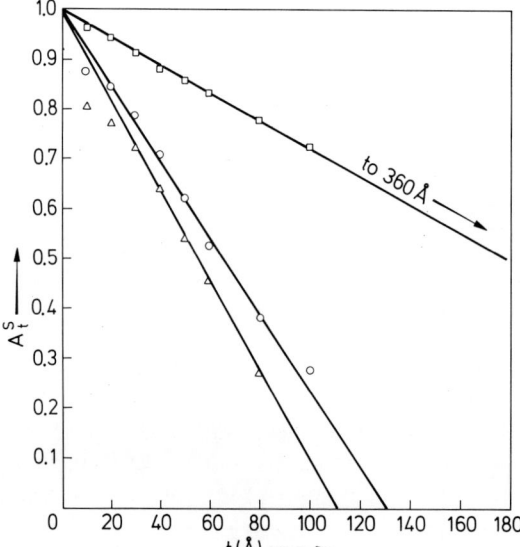

Fig. 3. The $A_t^s$ vs. t plot (Buchanan-Miller [10])

The variation of the strains, calculated from the slopes of the plots, like on the Fig. 2, shows a rapid decreasing with R and levels off to a constant value, as shows the Fig. 4.

Line broadening analysis was applied in different forms. As in most polymer systems it is difficult to observe even two orders of reflexion, using data from other reflexions (i.e. pseudo-orders) was investigated with the help of line broadening analysis.

Assuming, that both $\beta_S$ and $\beta_D$ arise from Cauchy profiles, the integral breadth $\beta$ of the resulting convolution product is

$$\beta = \beta_S + \beta_D \tag{15}$$

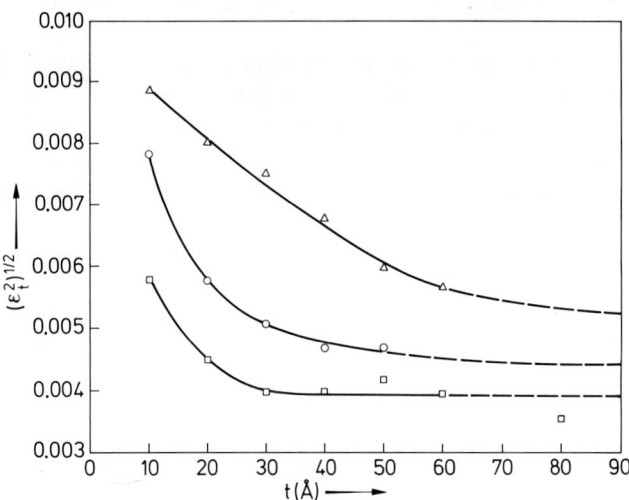

**Fig. 4.** $\varepsilon_t^2$ vs. t plot (Buchanan-Miller [10])

If $\beta_S$ and $\beta_D$ are assumed to result form Gaussian profiles, the integral breadth of the convolution product is

$$\beta^2 = \beta_S^2 + \beta_D^2 \tag{16}$$

One possible solution is, that size profile is Cauchy, while the distortion profile is Gaussian. This is the assumption proposed by Schoening, which was already discussed [8].

The separation with the Bonart, Hosemann, McCullogh method [6] was also investigated.

In all integral breadth methods the Scherrer equation is incorporated as size broadening component. This gives (with integral breadths) for the determination of crystallite size results in a weight-average size, in contrast to a number average size obtained from the Fourier transform method.

The results for all these assumptions were calculated. They are differring both in particle size and in the lattice distortion. Tha Cauchy assumption was found not good at all. The others are delivering particle size in the range of 100–300 Å, and distortion 0.2–5%, for the 3 investigated iPS samples. As relatively small amount of distortions are present, it was not possible to decide if paracrystalline distortions or microstrain distortions are present.

The paracrystalline distortions should broaden a line according to the square of the order of the reflexion and microstrain distortion should broaden a line, according to the first power of the order of reflexion.

A plot of the squared integral breadths against $n^2$ is given for three iPS samples in Fig. 5. Inaccuracies of the data make a categorical distinction impossible between the two possibilities. Minimum three orders of reflexion are required in order to distinguish between the two theories.

It was found, that the best results are given (after a Stokes deconvolution) by the Fourier transform method.

The size and distorsion effects could be separated for a polymer system. Instrumental requirements are very severe: precise line-profile data must be collected over large regions of reciprocal space, an accurate correction for background scattering is necessary and a reliable method to resolve the overlapping peaks must be found. In order to obtain the maximum available amount of information, the full Fourier transform method should be applied whenever possible.

The effects of lattice parameter fluctuation was included in the $A_t^D(n)$ expression by Takahashi (11). According this,

$$A_t^D(n) = \exp\{-2\pi^2 n^2 (\langle Z_t^2 \rangle + \langle e^2 \rangle)\} \tag{17}$$

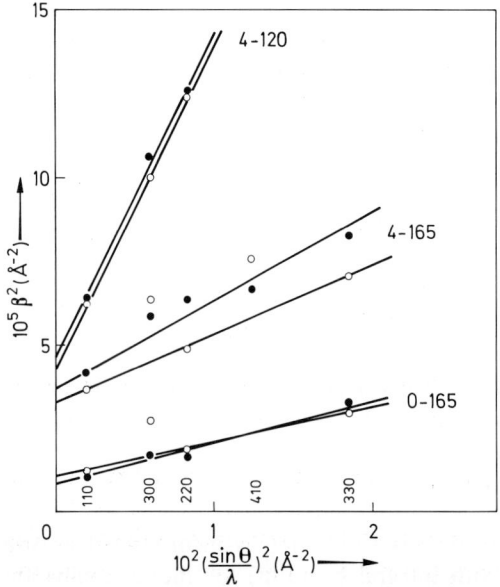

**Fig. 5.** Squared integralbreadths as a function of $(\sin \theta/\lambda)^2$ ○ instrumental correction by Fourier deconvolution ● instrumental correction according to the assumption, that all three profiles are Gaussian (Buchanan-Miller [10])

in which $e^2$ is the relative mean-square fluctuation of d among the crystals. $e$ is supposed to be independent of t. At the end of this chapter we will dicuss, that if such kind of fluctuation exist, the Warren-Averbach plot could not distinguish the particle size and the distortion caused by the fluctuation of d among the crystals.

A summary of the use of Fourier coefficients of paracrystalline X-ray diffraction was presented by Kulshreshtha and Dweltz [12]. They attempted to extend the Warren-Averbach technique of the line profile analysis by Fourier coefficients for paracrystals. The earlier results of Katayama [5] were reinterpreted. The following equation was used to get the lattice distortion and the crystalline particle size:

$$\ln A_t(n) = -\left[\frac{2\pi^2 g^2 n^2}{d} + \frac{1}{\bar{D}}\right] R \tag{18}$$

where g is the paracrystalline distortion, the average relative square dislocation of the plane:

$$g^2 = \Delta^2/d^2 \tag{19}$$

where $\Delta^2$ is the variance of the interplanar spacing

$$\Delta^2 = \langle d^2 \rangle - \langle d \rangle^2$$

The quantity $g^2$, the reduced variance of the interplanar separation vector should not be confused with $\langle e^2 \rangle$, the reduced variance of the lattice parameter $d = \langle d \rangle$ in each crystal within the sample.

From the two slopes of a plot $\ln A_t(n)$ vs. R, the g and $\bar{D}$ values could be calculated for the two PE samples. They received for $\bar{D}$ values 205 and 295 Å, and for g values 2.8 and 2.6% respectively for the PE samples studied by Katayama at $-196$ °C and 20 °C.

A study of deformed semicrystalline polymer was reported by Wecker, Cohen and Davidson [13]. The subject was polytetrafluorethylene and Fourier method was used to analyse the size changes and the internal lattice distortion. Three orders of the $10\bar{1}0$ reflexion was examined. The Warren-Averbach plot of a Teflon resin is reproduced in Fig. 6. This plot was used for the determination of the particle size and for the distortions. The intercepts at $n = 0$ are the $A_t^S$ values, the particle size Fourier coefficients.

From each slope a root mean square distortion was computed. The value of these data is decreasing with R: it is 11.3 at R = 10 and 2.9 at R = 150 in $\langle \varepsilon_R^2 \rangle^{0.5} \cdot 10^3$ units.

The integral widths of diffraction peaks are proportional to $n^2$ in the case of paracrystalline defects and to n in the case of microstrains. Schönfeld, Wilke, Höhne and Hosemann [14] proposed a method for the determination of the crystallite size and for lattice defects from the integral widths. To distinguish between microstrains and paracrystalline defects, at least 3 order of reflexions should be plotted in a $\beta$-$s^2$ diagram. Vogel, Haase and Hosemann [15] used both reflex linewidth and Fourier method for distinguish microstrains and paracrystalline distortions. The

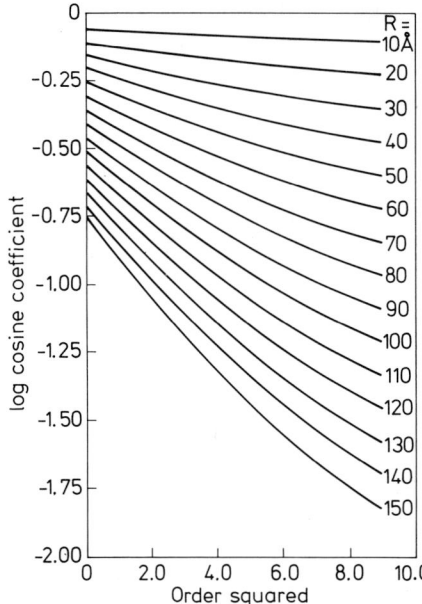

**Fig. 6.** Logarithm of the Fourier cosine coefficients vs. square of the order t. Renormalised coefficients. (Wecker, Cohen and Davidson [13])

prinziple is, that dividing the Fourier coefficients of two orders of reflexions, the crystalline particle coefficient term falls out, only the n depending terms are present:

$$\frac{A_t(n)}{A_t(n+1)} = \frac{A_t^S \cdot A_t^D(n)}{A_t^S \cdot A_t^D(n+1)} = \frac{A_t^D(n)}{A_t^D(n+1)} \tag{20}$$

From the plot of $\ln(A_t^D(n)/A_t^D(n+1))$ vs. t in the case of a paracrystalline distortion, the g value could be calculated:

$$\ln \frac{A_t^D(n)}{A_t^D(n+1)} = 2\pi^2(2n+1) g^2 t \tag{21}$$

In the case of a microstrain:

$$\ln \frac{A_t^D(n)}{A_t^D(n+1)} = 2\pi^2(2n+1) \langle \varepsilon^2 \rangle t^2 \tag{22}$$

The Eq. (22) is valid only in the case of microstrains having Gaussian distribution. If both paracrystalline and microstrains with Gaussian distribution are present,

$$\ln \frac{A_t^D(n)}{A_t^D(n+1)} = 2\pi^2(2n+1)(g^2 t + \langle \varepsilon^2 \rangle t^2) \tag{23}$$

The method was applied to different samples, here we give the results for a linear, from melt crystallized PE. In Fig. 7., the 200, 400 and the 110, 220 reflexes are

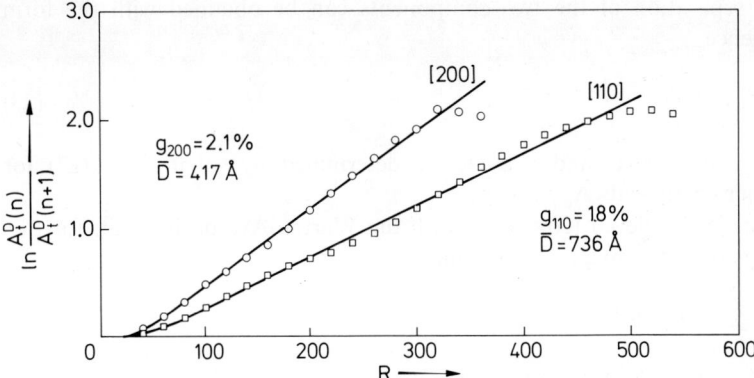

**Fig. 7.** The plot of ln $(A_t(n)/A_t(n+1))$ versus n. Linear PE, crystallized from the melt. (Vogel, Haase and Hosemann [15])

demonstrated, showing linearity in t dependence. The abscissa is in $R = t \cdot d$ units calculated.

The separation of the particle size and microstrain components in the Fourier coefficients of a single diffraction profile was proposed by Gangulee [16].

He proposed functional forms for the dependence of the average microstrain on the averaging distance.

The particle size component $A_t^S$ is expressed for small values of t as

$$A_t^S = 1 - tx$$

where $x = \delta/\bar{D}$, $\delta$ is a distance, whose magnitude is inversely proportional to the Fourier period, here 20 Å. $R = t \cdot x$ the real averaging distance normal to the hkl planes.

The microstrain term for small values of t may be written as:

$$A_t^D(n) = 1 - t^2 K \langle \varepsilon_t^2 \rangle$$

where $K = 2\pi^2 \delta^2/d$

$\langle \varepsilon_t^2 \rangle$, the average squared microstrain, is averaged over a distance t. $\delta$, d is the spacing between the hkl planes.

Let

$$Y_t = K \cdot \langle \varepsilon_t^2 \rangle \tag{26}$$

and then the low harmonic cosine Fourier coefficients from an hkl reflexion can be written as

$$A_t(n) = (1 - tx)(1 - t^2 Y_t) \tag{27}$$

In principle, the Equation can be solved for x if the functional form of $Y_t$ is known, thus separating the particle size and the microstrain terms.

Acceptable separation of the two components can be obtained with the form $\langle \varepsilon_t^2 \rangle = C/t$, hence

$$Y_t = KC/t \tag{28}$$

where C is a constant. C and x could be determined by solving Eq. (27) for any two Fourier coefficients $A_{t,1}$ and $A_{t,2}$.

Better results are achieved (compared with the Warren-Averbach result, for two harmonic reflexions) by using an approximation

$$\langle \varepsilon_t^2 \rangle = C_1 + (C_2/t) \tag{29}$$

$$Y_t = KC_1 + (KC_2/t) \tag{30}$$

The values of x, $C_1$ and $C_2$ can be obtained by solving Eq. (27) for three coefficients $A_{t,1}$, $A_{t,2}$, $A_{t,3}$. Solutions for t, 1; t, 2; t, 3 = 3, 4, 5 gave the best fit with the Warren-Averbach method. (Table 2.)

It seems, that adequate separation of the particle size and microstrain components in the cosine Fourier coefficients of a single diffraction profile is found, if an appropriate microstrain function is assumed.

Mignot and Rondot proposed a 2nd degree polynomial approximation of the $A_t(n)$ Fourier coefficients [17]. This P(t) polynom could be fitted by the least square method to $A_t(n)$. The results are quite near to the Warren-Averbach (two line) results. (Table 2, last column.)

The Voigt function was used for the analysis of the breadths of diffraction and spectral lines by Langford [18]. A single line profile could be described with a Cauchy (Lorenzian) or with a Gaussian function. The convolution of these functions is known as a Voigt function, of which the Cauchy and the Gaussian curves are the limiting cases.

A symmetrical line profile could be described in terms of three parameters. These are: 1. peak height I(0), 2. area, A and 3. full width at half of the maximum intensity (the half width), 2w.

**Table 2.** The microstrains corresponding to the particle size obtained with different approximations. (Gangulee [16], Mignot, Rondot [17])

| t | Warren-Averbach | Gangulee I | Gangulee II | Mignot-Rondot |
|---|---|---|---|---|
|   | $\bar{D} = 506$ Å | $\bar{D} = 499$ Å | $\bar{D} = 506$ Å | $\bar{D} = 509$ Å |
|   | | $\langle \varepsilon_t^2 \rangle^{1/2} \cdot 10^{-3}$ | | |
| 1 | 2.05 | 1.85 | 1.96 | 2.32 |
| 2 | 1.49 | 1.42 | 1.49 | 1.64 |
| 3 | 1.24 | 1.20 | 1.26 | 1.34 |
| 4 | 1.09 | 1.04 | 1.09 | 1.16 |
| 5 | 0.99 | 0.94 | 0.98 | 1.04 |

1. and 2. may be combined to give the integral breadth, β where

$$\beta = \frac{A}{I(0)} \tag{31}$$

All three parameters are included in the ratio $2w/\beta$, and any symmetrical line profile could be characterized by this ratio, which was named "form factor".

The Cauchy function is used in the form

$$I_C = I_C(0) \frac{w_C^2}{w_C^2 + x^2} \tag{32}$$

The area of the function is

$$A_C = \pi w_C I_C(0) \tag{33}$$

and the integral breadth:

$$\beta_C = \pi w_C \tag{34}$$

The form factor $2w/\beta$ for a Cauchy function is $2/\pi = 0.63662$.

The Gaussian function is used in the form

$$I_G = I_G(0) \exp(-\pi x^2/\beta_G) \tag{35}$$

the area is

$$A_G = I_G(0) \beta_G \tag{36}$$

and the half-width

$$2w_G = 2\beta_G (\ln 2)^{1/2}/\pi^{1/2} \tag{37}$$

The integral breadth is

$$\beta_G^2 = \frac{\pi}{4 \ln 2} (2w_G)^2 \tag{38}$$

The form factor for a Gaussian function is $2(\ln 2)^{1/2}/\pi^{1/2} = 0.93949$.

The form factor has intermediate values for a Voigt function. Providing, that peak heights and areas are known, Voigt curves for any combination of Cauchy and Gaussian function can be generated. (If only one is known, the other may be obtained from the half width or integral breadth.)

The Voigt profile in terms of the parameters defining the constituent Cauchy and Gauss curves:

$$I(x) = \text{Re} \left\{ \beta_C I_C(0) I_G(0) \, \omega \left[ \frac{\pi^{1/2} x}{\beta_G} + ik \right] \right\} \tag{39}$$

where ω is the complex error function, defined as

$$\omega(z) = \exp(-z^2)\left[1 + \frac{i2}{\pi^{1/2}} \int_0^z \exp(t^2)\, dt\right] \quad (40)$$

The complex error function can be obtained from standard tables, such are given in Abramowitz and Stegun [19], for example.

k is a factor proportional to the ratio of the Cauchy and Gaussian integral breadths, given by

$$k = \beta_C/\pi^{1/2}\beta_G \quad (41)$$

The integral breadth is

$$\beta = \frac{\beta_G \exp(-k^2)}{1 - \mathrm{erfc}(k)} \quad (42)$$

The full width at half height of the Voigt function, 2w is obtained from

$$\mathrm{Re}\left\{\omega\left[\frac{\pi^{1/2}w}{\beta_G} + ik\right]\right\} = \frac{1}{2}\omega[ik] = \beta_G/2\beta \quad (43)$$

which approximates to

$$(2w)^2 \sim \frac{4\beta_G^2}{\pi}(1+k^2) \sim \frac{(2w_G)^2}{\ln 2} + (2w_C)^2 \quad (44)$$

The dimensionless quantities $\beta/\beta_G$ (Eq. (42)) and $2w/\beta_G$ (from Eq. (43)) can be calculated as functions of k.

In practial applications, the parameters defining the consistent profiles are to be obtained from a broadened line. If the profile is assumed to be a Voigt curve, Eq. (43) can in principle be solved to find k and hence $\beta_G$ and $\beta_C$ for measured values of 2w and β. Numerical and graphical methods are proposed for the solution. For any value of k, $\beta_G/\beta$ can be calculated from Eq. (42). The corresponding values of $\pi^{1/2}w/\beta_G$ and hence $2w/\beta$ are then obtained by interpolation from the same tables, using the condition, that the real part of $\omega[\pi^{1/2}w/\beta_C + ik]$ is equal to $\beta_G/2\beta$.

The $2w/\beta$, k and $\beta_G/\beta$ values are listed in the Table 2 of [18]. From this Table a plot was made to calculate the $\beta_G/\beta$ values from the measured $2w/\beta$, which is given in Fig. 8.

The Voigt function was used in a single line method for the analysis of X-ray line broadening by de Keijser, Langford, Mittenmeijer and Vogels [20]. According to Langford [18], they were calculating the $2w/\beta$ and $\beta_G/\beta$ values. Instead of graphical methods or interpolation from tables, empirical formulae of high accuracy were used. In single line analysis it is assumed, that the Cauchy component is due to the

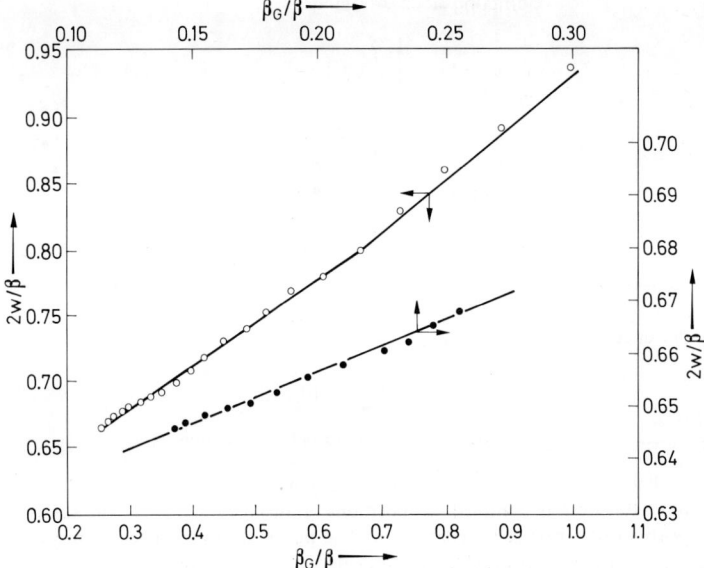

**Fig. 8.** The $\beta_G/\beta$ values plotted against $2w/\beta$, according to the results of Langford [18], Table 2

crystallite size and that the Gaussian contribution arises from strain. The crystallite size D is given by, according to Eqn. 8 and 9:

$$D = \frac{\lambda}{\beta_C \cos \theta} \tag{45}$$

and the strain e by

$$e = \frac{\beta_G}{4 \operatorname{tg} \theta} \tag{46}$$

where $\beta$ is measured on a $2\theta$ scale, according to Halder and Wagner [21].

An overview of the determination of crystallite size and lattice distortions through X-ray diffraction line profile analysis was given by Delhez, de Keijser and Mittemeijer [22]. Attention was paid mainly to the Fourier analysis of line profiles on the basis of the theory of Warren and Averbach and to the analysis in terms of breadth parameters on the basis of Voigt functions, but many other methods were treated.

Especially interesting are their remarks on the effects of sampling and truncation of the profile on the Fourier coefficients. In Fig. 9. a demonstration is given, what the sampling of a line profile means to the Fourier coefficients.

The effect of a bad background determination causes a truncation of the line profile. If the intensity at the point of truncation should be set equal to zero, the profile, indicated by · is obtained. For such profiles, a hook effect and serious distortions occur in Fourier space, as can be seen on Fig. 10.

The truncation and sampling in real space cause a hook effect and aliasing respectively in Fourier space. The hook effect is a negative curvature in the

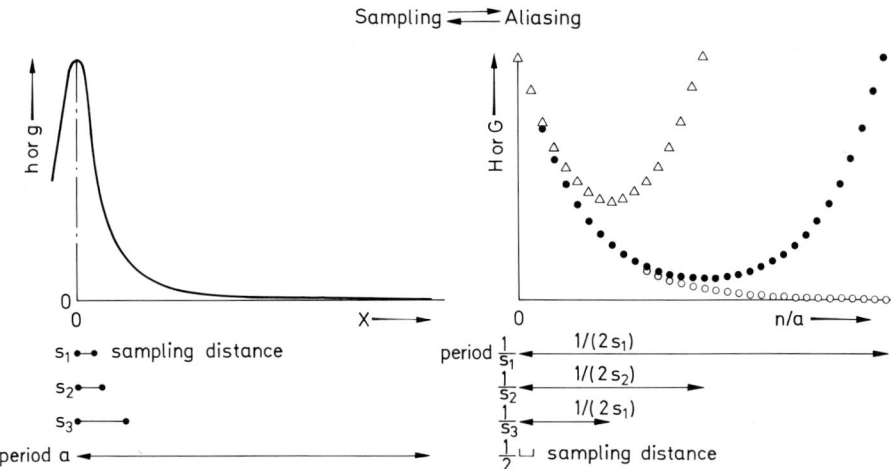

**Fig. 9.** The sampling of a line profile and its consequence: the aliasing of the Fourier coefficients. In the example a Cauchy function is sampled at positions a distance $s_1$, or $s_2$ or $s_3$ apart corresponding in Fourier space to a period $1/s_1$, $1/s_2$, $1/s_3$. The period $a$ in real space corresponds to the sampling distance $1/a$ in Fourier space. (Delhez, de Keijser and Mittemeijer [22])

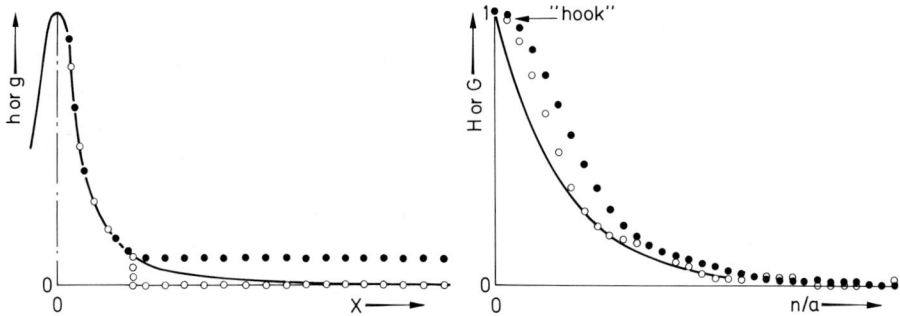

**Fig. 10.** The effect of truncation of the line profile on the Fourier coefficients. In the example a Cauchy function is truncated as indicated by ●. To approach a practical situation the intensity at the point of truncation should be set equal to zero (background estimated too high) and the profile as indicated by o is obtained. For both profiles ● and ○ apart from the hook effects, serious distortions occur in the Fourier space. (Delhez, de Keijser and Mittemeijer [22])

experimental $A_t(n)$ and $A_t^S$ curves near to $t = 0$. It can be shown, that a hook effect in the curve $A_t^S$ versus t is impossible. However a hook effect in the curve of $A_t(n)$ versus t is not necessarily an artefact of the measurement, but it can be due to the strain present in the sample. It is therefore dangerous to correct the $A_t(n)$ curve for an observed hook effect.

A summary about the Fourier analysis of polymer X-ray diffraction patterns was given by Crist and Cohen [23]. They underlined that considerable confusion exists regarding interpretation of the line shapes of diffraction peaks: there is no accepted method for analyzing the peak profiles of the diffraction pattern. They say that instrumental broadening can be quantitatively corrected for by the method of Stokes [1].

After this step the Fourier coefficients are available, — it is these coefficients which may be used directly as Warren and Averbach suggested [2] to separate the effects of size and disorder. This Fourier analysis has been fruitfully empolyed in the study of metals and alloys, although relatively few applications have been made to polymers. This situation exist partly because the disorder or distortion effects observed by Fourier analysis have been described by the somewhat ambiguous term "microstrains".

They explain, that paracrystallinity is just one type of disorder which may be inferred from the general Fourier analysis. They want to point out that dislocations within crystals give rise to peak broadening which is indistinguishable from paracrystallinity.

The author of the present article has the opinion, that the strain might be more important in the case of polymers, than in metals. Extended amorphous chains are connecting the lamellas, causing stress-cracking and ageing in polymers. These extended amorphous chains are not considered up till now as the reason for strain and for line broadening.

Some people are regarding these oriented amorpous chains as a third phase: Raman spectroscopy is able to measure crystallinity and amorphycity of polymers. Some percent is missing to the 100%, this is supposed to be the third phase, the interlamella chains, extended in their form, which are connecting the lamellas. [24,25]

In metals such interconnecting amorphous chain does not exist. As a matter of fact, the basic difference between small and large molecular systems seems to be not only in paracrystallinity, which causes second kind of distortions, but in the extended chains existing in the amorphous phase. Single chains of course are not visible with our present methods, but their effect could be measured and X-ray line shape analysis seems to be an excellent method for doing it. X-ray, Raman and electronmicroscope studies could together clear the picture.

Crist and Cohen calculated the size and distortion from the data published by Katayama [5]. This recalculation was made by the 7 methods, given in the earlier literature. It seems, that with the exception of Katayama's method, all of the Fourier separations give qualitatively similar results, even the integral breadth processes are quite good.

At the end of their article, they stated, that single line methods, like the process proposed by Mignot and Rondot [17], could be used for single line profile analysis. Least squares methods for Fourier analysis are available to replace the Fourier inversions [26].

Schmidt [27] has pointed out, that in heat treated, undrawn PE, the (002) direction has no distortion and he was able to calculate the $D_{002}$ values from the integral breadth of the peaks. The values were in the range of 190 to 350 Å, depending on the temperature of the heat treatment.

The method of Warren and Averbach was used for one single line by Krenzer and Ruland, for studying the change of the profile as a function of the temperature [28].

From the Eq. 7., the $A_t^D(n)$ contains information on the lattice distortions:

$$A_t^D(n) = \exp \frac{2\pi^2 \sigma_{hkl}^2(t)}{d_{hkl}^2} \qquad (47)$$

where $\sigma_{hkl}^2$ is the variance of the distribution of distances between any two net planes t spacings apart and $d_{hkl}$ is the Bragg spacing of the index hkl. The variation of

$\sigma_{hkl}^2$ with t contains the information of the type of lattice distortions. In the case of a paracrystalline disorder $\sigma_{hkl}^2$ is proportional to t, in the case of a strain distribution, $\sigma_{kl}^2$ is proportional to $t^2$:

$$\sigma_{hkl}^2 = t^2 d_{hkl}^2 \Delta\varepsilon^2 \qquad (48)$$

where $\Delta\varepsilon$ is the strain variance.

Assuming, that going down from the room temperature the crystallite size is not changing with the temperature, the Fourier transforms of two line profiles of the same index measured at two different temperatures $T_1$ and $T_2$, the ratio of their cos Fourier coefficients, $A_t(T_1, n)/A_t(T_2, n)$ is given by

$$\frac{A_t(T_1, n)}{A_t(T_2, n)} = e^{-\frac{2\pi^2}{d_{hkl}^2}[\sigma_{hkl}^2(T_1, t) - \sigma_{hkl}^2(T_2, t)]} \qquad (49)$$

hence

$$\Delta\sigma^2 = \sigma_{hkl}^2(T_1, n) - \sigma_{hkl}^2(T_2, n) = \frac{d_{hkl}^2}{2\pi^2} \ln \frac{A_t(T_2, n)}{A_t(T_1, n)} \qquad (50)$$

The plot of the half peak width (FWHI) in s units as a function of the temperature shows, that the increase of the width with decreasing temperature is not monotonic: inflexion points are at about 259 K and 134 K, which corresponds to the temperature of the β and γ relaxations in polyethylene.

To summarize this chapter on the calculations from the structural broadening, we can say that a very broad literature of the distortion determination has been published. The definitions are in this article a little uniformized, but the author had difficulties to do it. Still today do not exist a clear way for the determination of the different kind of distortions, but the distinction between the distortion and the particle size Fourier coefficients is fairly developed. Still some problems we have to mention.

If there is a fluctuation in the average strain of each particle, as it was considered by Takahashi [11] and the Fourier coefficient $A_t^D$ is consisting of two factors, one of which due to the fluctuation of the average strain of each particle, named $A_t^e$ and the other is due to the inhomogeneous strain, named $A_t^d$, the particle size and the average strain could not be distinguished from the Warren-Averbach plot.

Already Takahashi was pointing out, that in the case, if there are m random variables with their m distribution functions, the distribution function of the sum is given by the convolution of the m distribution functions and the Fourier transform of the distribution functions is given by the product of the Fourier transforms of the distribution functions.

Thus the distortion term is the product of the two kinds of distribution:

$$A_t^D(n) = A_t^e \cdot A^d(n) \qquad (51)$$

from which $A_t^e$ is not dependent on n.

The Eq. (10) has to be written in the form:

$$A_t(n) = A_t^S \cdot A_t^e \cdot A_t^d(n) \qquad (52)$$

The Warren-Averbach method is based on the n dependence of the Fourier coefficients: particle size coefficient is supposed to be independent of n, while the distortion was supposed to be n dependent. Now it seems, that the Warren-Averbach plot delivers the product of the not n dependent particle size and average strain coefficients and the n dependent inhomogeneous strain. As the $A_t^s$ and $A_t^e$ are both independent of n, it seems no direct way to see, what part belongs from the intercept of a Warren-Averbach plot to the particle size and what to the average strain.

For a single line, the line shape analysis is still difficult. The Gangulee [16] — Mignot-Rondot [17] method is promising a possibility for such measurements.

The Langford [18], De Keijser and cow. [19] method is useful in the case, if the particle size gives Cauchy and the distortion gives a Gaussian function line shape. The author made with this assumptions synthetic Voigt functions and from those, the original, known Cauchy and Gaussian components were supposed to get back.

Such a synthetised Voigt curve is shown with its components in Fig. 11.

As a matter of fact, we had difficulties with getting back the original Cauchy components. The reason of the difficulties is probably that the Cauchy and the Voigt function has very broad tails and the integration in order to have the integral breadths have to be carried out on an unusually broad scale.

Many authors are supposing, that the particle size distribution gives a Cauchy-type peak profile. In this connection we see some limitations. As we shall see later on, the second derivative of the particle size Fourier coefficients vs. t gives the particle size distribution. As negative particle size does not exist, the particle size Fourier coefficients vs. t could not have a positive curvature, not even an inflexion point as the second derivative of an inflexion point is zero, running on one side into the negative range.

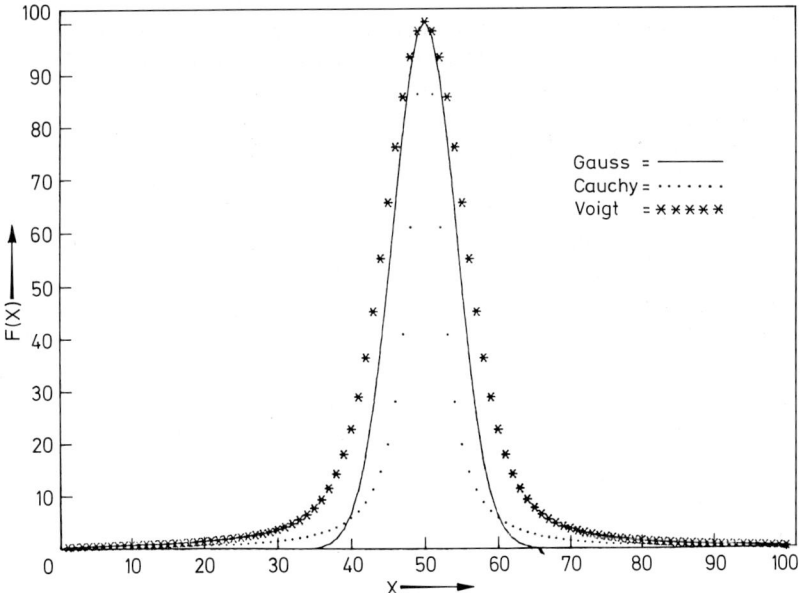

Fig. 11. The convolution of a Cauchy and a Gaussian curve, resulting a Voigt peak

In the case we have the distortions in a shape of a Gaussian function and we have received from the Langford [18] — De Keijser [19] method or from other sources the Gaussian ratio of the structurally broadened peak, there is a possibility to calculate the particle size and the distortion Fourier coefficients by the deconvolution method. In such a case h(x) is the structurally broadened peak, g(x) is a Gaussian function with known ratio of the 2w half peak breadth and f(x), the result of the (second) Stokes deconvolution is the particle size broadened peak. There is no assumption about the peak profile of the particle size broadening. This second deconvolution could be important if only a single peak could be registered.

## 3 Calculations from the Distortion and from the Particle Size Fourier Coefficients

If we have the separated $A_t^D(n)$ and $A_t^S$ cosine Fourier coefficients, the distribution of the distorsion and crystalline particle size could be calculated.

As for the distortion coefficients, a plot of $A_t^D(n)$ versus R for different crystallographic directions gives a general picture about the distortions in different directions. From such a plot McKeehan and Warren [29] were calculating the strain distribution for cold worked tungsten.

$p(\varepsilon_F)$, the strain distribution function can be obtained for a given value of R from the Eq. (53):

$$p(\varepsilon_R) = \int_0^\infty A_t^D(n) (l_0) \cos (2\pi l_0 R \varepsilon_R / a) \, dl_0 \tag{53}$$

where $l_0^2 = h^2 + k^2 + l^2$ and $a = d \cdot l_0$.

For some selected value of R, the strain distribution function can be calculated.

For polymers, this method was applied by Buchanan and Miller [10]. Their results were presented in Fig. 4.

The crystalline particle size distribution was first calculated by Bertaut [30,31,32], who prooved, that the second derivative of the $A_t^S$ curve vs. R gives directly the particle size distribution. His work was connected with metals.

The method was modified and used to carbon black crystallites size distribution determination by Doi [33]. Polyethylene crystallite size distribution and lattice distortion determination was reported by Yoda and cow. [34,35,36]. In their works the crystalline particle size analysis has been often affected by meaningless oscillations, particularly in the small size region [37].

A method, supressing spurious oscillations due to the truncation effect was proposed by Odajima et al. [37]. They were showing, that for small particle sizes the distribution curve is less reliable, as it is mainly affected by the low intensity in the far tails of the diffraction profile, where the relative error is very high. Their method was applied to polyoxymethylene.

A computer program was developed for the determination of the crystalline particle size distribution by Bodor and Füzes [38,39].

The results of an investigation of a linear PE, Marlex 6050 are presented. The sample was melted for 2 hours at 160 °C and annealed for 2 days at 125 °C. The 110 and the 220 peaks, after a Stokes deconvolution were extrapolated, according to Warren and Averbach [2]. The result is shown in Fig. 12. The $A_t^S$ cosine Fourier versus t are plotted in Fig. 13.

The number size distribution of the crystalline particles, calculated as

$$\frac{d^2 A_t^S}{dt^2 A_{t=0}^S} \tag{54}$$

is shown in Fig. 14. The number average crystalline particle size,

$$\bar{D}_n = \frac{\sum_t A_t^S \cdot R(t)}{\sum_t A_t^S} \tag{55}$$

was found $\bar{D}_n = 11.6$ nm.

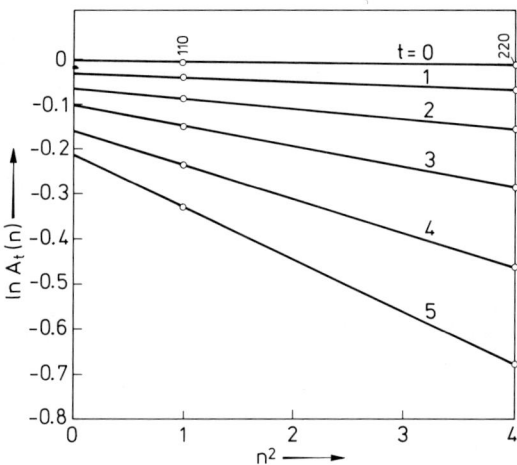

**Fig. 12.** The Warren-Averbach plot of Marlex 6050, melted at 160 °C and crystallized at 125 °C for two days. Reflexions 110 and 220

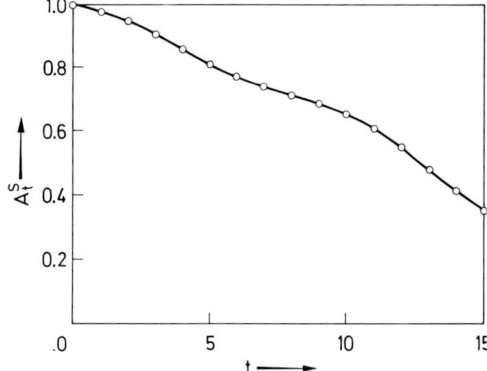

**Fig. 13.** The Fourier coefficients of the particle size broadening $A_t^S$ vs. t from the Fig. 12

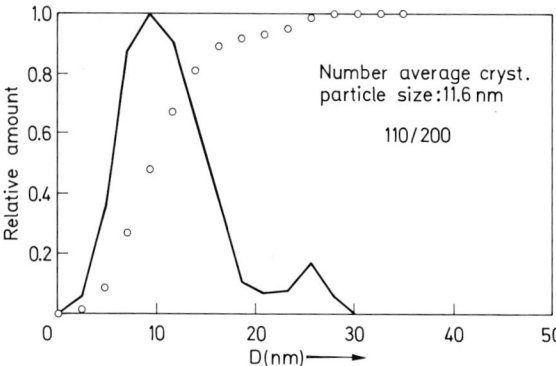

Fig. 14. The number distribution of the crystalline particle size, Marlex 6050

The diameter size distribution of the crystalline particle size, given by

$$\frac{d^2 A_t^S}{dt^2} \frac{R(t)}{A_{t=0}^S R(0)} \tag{56}$$

is plotted in Fig. 15.
The diameter average crystalline particle size,

$$\bar{D}_d = \frac{\sum_t A_t^S \cdot R(n)^2}{\sum_t A_t^S \cdot R(n)} \tag{57}$$

was measured $\bar{D}_d = 14.0$ nm.

As for the 200, 400 reflexions, the Warren-Averbach plot is presented in Fig. 16, the $A_t^S$ cosine Fourier coefficients versus t in Fig. 17. The number size crystalline particle distribution is given in Fig. 18, with an $\check{D}_n = 16.8$ nm, the diameter size distribution is shown in Fig. 19, the diameter average was $\bar{D}_d = 17.5$ nm.

The computer program was extended to polypropylene and to polyamide, too.

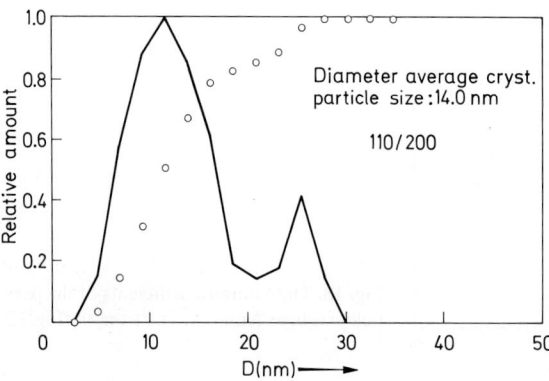

Fig. 15. The diameter distribution of the crystalline particle size, Marlex 6050

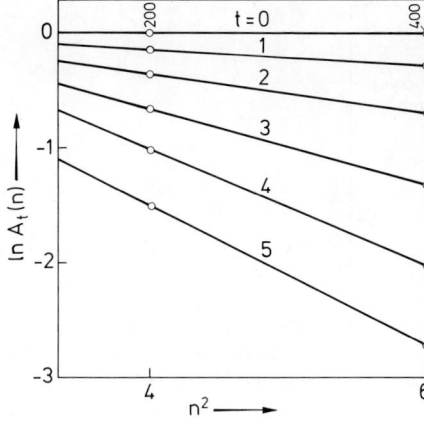

**Fig. 16.** The Warren-Averbach plot of Marlex 6050. Reflexions 200 and 400

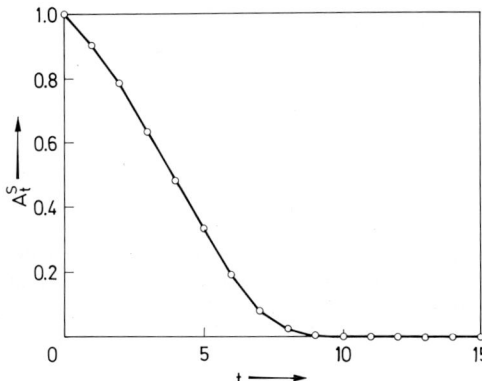

**Fig. 17.** The $A_t^S$ Fourier coefficients vs. t from the Fig. 16

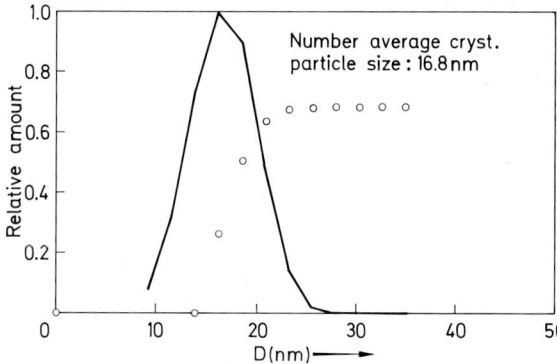

**Fig. 18.** The number distribution of the crystalline particle size from Fig. 17. Marlex 6050

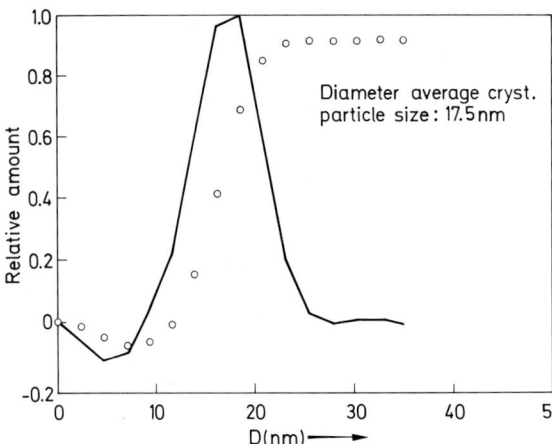

**Fig. 19.** The diameter distribution of the crystalline particle size from Fig. 17. Marlex 6050

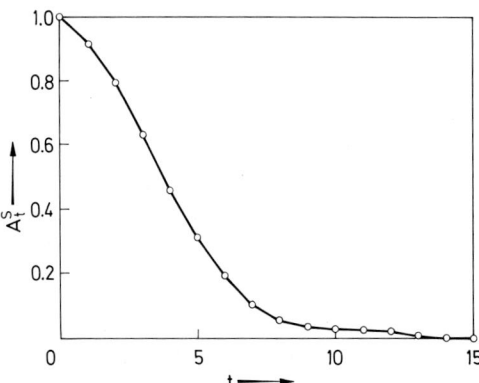

**Fig. 20.** The $A_t^S$ crystalline particle size cosine Fourier coefficients of the 200/400 reflexions for a Linear Low Density Polyethylene, melted at 145 °C, crystallized at 120 °C for two days

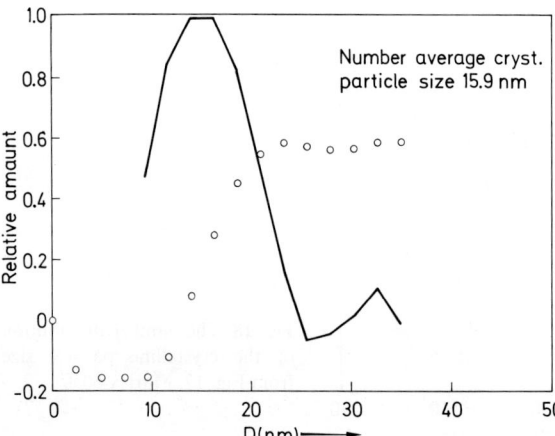

**Fig. 21.** The number distribution of the crystalline particle size from the Fig. 20. (LLDPE.)

In the past 5 years, most of our problem was connected with the proper handling of the amorphous background and with the overlapping of the peaks.

An other investigation is presented, where the crystalline particle size is clearly bimodal. The sample was a Linear Low Density Polyethylene (LLDPE) melted for 2.5 hours at 145 °C and crystallized at 120 °C for two days. The $A_t^s$ cosine Fourier coefficients versus t are plotted in Fig. 20, the number and the diameter distributions are shown in Figs. 21. and 22.

The distortion and crystalline particle size distributions are important for the better characterization of polymer structures. The strains and distortions are directly connected with the mechanical properties. The crystalline particle size distribution determinations are now in the stage of introduction into the practical applications.

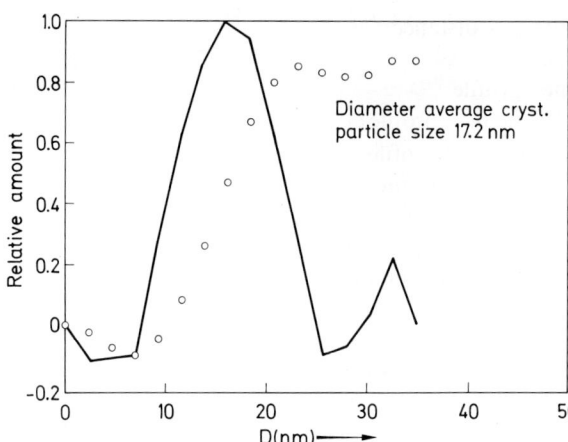

**Fig. 22.** The diameter distribution of the crystalline particle size from the Fig. 20. (LLDPE.)

## 4 Main Symbols Used in the Text

Some symbols in this article are used as in the general X-ray literature and not as in the cited articles.

d   is the Bragg spacing
D   crystalline particle size
L   long period
n   order of the reflexion
t   harmonic number of the Fourier coefficient

$s = \dfrac{2 \sin \theta}{\lambda}$    $\theta$ is the Bragg angle

$\lambda$   is the wavelenght of the radiation

$R = ta_3$, where

$1/a_3 = \dfrac{2 \sin \theta_1}{\lambda} - \dfrac{2 \sin \theta_2}{\lambda}$

$\theta_1$ and $\theta_2$ are the positions of tails of the peak. In some cases $R = td$ or $R = tx$, where d is a Bragg spacing and x is an arbitrary distance.

e = $\Delta d_{hkl}/d_{hkl}$, the lattice parameter fluctuation ($e^2$ is the reduced variance of the lattice parameter $d = \langle d \rangle$)

g is the paracrystallinty parameter, the reduced variance of the interplanar separation vector, defined by

$$g^2 = \frac{\Delta^2}{\langle d \rangle^2}$$

where

$$\Delta^2 = \langle d^2 \rangle - \langle d \rangle^2$$

$\langle \varepsilon^2 \rangle$ is the mean square microstrain

$\langle Z_t^2 \rangle$ is the root mean square relative displacement

$$\langle Z_t^2 \rangle = \langle \varepsilon_R^2 \rangle R^2/d^2_{n=1}$$

or

$$\langle Z_t^2 \rangle = \langle \varepsilon_R^2 \rangle t^2$$

h(x)  is the experimental profile in x distance
g(x)  is the instrumental profile
f(x)  is the structural broadened profile
H(t)  the Fourier coefficients of the h(x) profile
G(t)  the Fourier coefficients of the g(x) profile
F(t)  the Fourier coefficients of the f(x) profile
$A_t(n)$  the cosine part of the F(t)
$A_t^S$  the cosine part of the particle size broadening
$A_t^D(n)$  the cosine part of the distortion broadening
β  integral breadth
$β_S$  integral breadth of the particle size broadening
$β_D$  integral breadth of the distortion broadening

## 5 Summary

An overview is given on the methods of X-ray line shape analysis as a means for characterization of crystalline polymers. The possibilities for the determination of the distortion and the crystalline particle size parts of the wide angle peak broadening are discussed.

Line breadth methods and Fourier methods are outlined for the cases, where at least two order of reflexions are available and special interest is given to the cases, where only one single peak could be registered.

The possibilities of the distribution determinations for distortion and for the crystalline particle size are explained.

*Keywords*: X-ray line shape analysis
stress, distortion and their distribution determination
crystalline particle size distribution, deconvolution

## 6 References

1. Stokes, A. R.: Proc. Philos. Soc. London *A61*, 382 (1948)
2. Warren, B. E., Averbach, B. L.: J. Appl. Phys., **23**, 497 (1952)
3. Wilson, A. J. C.: X-Ray Optics (Methuen and Co, London, 1949) p. 5

4. Klug, H. P., Alexander, L. E.: X-Ray Diffraction Procedures (John Wiley, New York 1954) Chap. 9.
5. Katayama, K.: J. Phys. Soc. Japan *16*, 462 (1961)
6. Bonart, R., Hosemann, R., McCullogh, R. L.: Polymer *4* 199–210 (1963)
7. Thielke, H. G., Billmeyer, F. W.: J. Polym. Sci. *A2*, 2947–2950 (1964)
8. Schoening, F. R. L.: Acta Cryst. *18* 975–976 (1965)
9. Wagner, C. N. J.: in Local Atomic Arrangements Studied by X-Ray Diffraction, Metallurgical Society Conf. *Vol. 36*, p. 217–269. Edited by Cohen J. B. and Hilliard J. E., Gordon and Breach, New York, 1965
10. Buchanan, D. R., Miller, R. L.: J. Appl. Phys. *37*, 4003–4012 (1966)
11. Takahashi, H.: J. Phys. Soc. Japan *27*, 708 (1969)
12. Kulshreshtha, A. K., Dweltz, N. W.: Acta Crystallogr. *A27*, 670 (1971)
13. Wecker, S. M., Cohen, J. B., Davidson, T.: J. Appl. Phys. *45* 4453–57 (1974).
14. Schönfeld, A., Wilke, W., Höhne, G., Hosemann, R.: Koll. Z. u. Z. Polym. *250*, 102 (1972)
15. Vogel, W., Haase, J., Hosemann, R.: Z. Naturforch *29a* 1152–58 (1974)
16. Gangulee, A.: J. Appl. Cryst. *7*, 434–439 (1974)
17. Mignot, J., Rondot, D.: Acta Metallurgica *23* 1321–24 (1975)
18. Langford, J. I.: J. Appl. Cryst *11*, 10–14 (1978)
19. Ambramowitz, M., Stegun, I. A. (1965) Handbook of Mathematical Functions. New York. Dower.
20. Dekeijser, Th. H., Langford, J. I., Mittemeijer, E. J., Vogels, B. P.: J. Appl. Cryst *15*, 308–314 (1982)
21. Halder, N. C., Wagner, C. N.: Acta Cryst. *20* 312–313 (1966)
22. Delhez, R., DeKeijser, Th. H., Mittemijer, E. J.: Fresenius Z. Anal. Chem. *312*, 1–16 (1982)
23. Crist, B., Cohen, J. B.: J. Polym. Sci. Polymer Physics Edition, *17*, 1001–1010 (1979)
24. Strobl, G. R., Hagedorn, W.: J. Polym. Sci. Polym. Phys. *16*, 1181–93 (1978)
25. Glotin, M., Mandelkern, I.: Colloid Polym. Sci. *260*, 182–192 (1982).
26. Kidron, A., De Angelis, R. J.: In Symposium on Computer Aided Engineering, Gladwell, G. M. L. Ed. Univ. of Waterloo, Canada 1971 pp. 285–297
27. Schmidt, W.: Dissertation, Berlin 1980
28. Krenzer, E., Ruland, W.: Colloid Polym. Sci. *259*, 405–412 (1981)
29. McKeehan, M., Warren, B. E.: J. Appl. Phys. *24*, 52–56 (1953)
30. Bertaut, E. F.: Comp. Rend. *228*, 187 (1949)
31. Bertaut, E. F.: Acta Cryst. *3*, 14 (1950)
32. Bertaut, E. F.: Acta Cryst. *5*, 117 (1952)
33. Doi, K.: Acta Cryst. *14*, 830 (1961)
34. Yoda, O., Doi, K., Tamura, N., Kuriyama, I.: J. Appl. Phys. *44*, 2211–2217 (1973)
35. Yoda, O., Tamura, N., Doi, K.: J. Materials Sci. *11*, 696–702 (1976)
36. Yoda, O., Kuriyama, I.: J. Polym. Sci. Phys. Ed. *15* 787–793 (1977)
37. Odajima, A., Noto, N., Yamane, S., Ishibashi, T.: Report on Progress in Polymer Physics in Japan *23*, 205–208 (1980)
38. Bodor, G., Füzes, L.: Crystalline Particle Size Distribution Determination. Conference on Diffraction Profile Analysis, Cracow, Poland 1978
39. Bodor, G., Füzes, L.: JUPAC Int. Symposium, Amherst, Mass. U.S.A. 1982. Proceedings p. 653.

H.-G. Zachmann (Editor)
Received Mai 28, 1984

# Use of Transmission Electron Microscopy to Obtain Quantitative Information About Polymers

I. G. Voigt-Martin
Institut für Physikalische Chemie, Universität Mainz, 6500 Mainz, FRG

*In the polymer field the electron microscope is frequently used as a tool for obtaining purely qualitative data, but only rarely to obtain quantitative information. In order to discuss the potential of this instrument for the latter purpose it is convenient to separate the discussion into the following sub-sections.*

1 Thickness Distribution in Linear PE Material . . . . . . . . . . . . . . . 196

2 Thickness Distribution in Co-Polymers . . . . . . . . . . . . . . . . . . 204

3 Evaluation of Molecular Tilt Angle . . . . . . . . . . . . . . . . . . . . 209

4 Radial Distribution Analysis . . . . . . . . . . . . . . . . . . . . . . . 212

5 Analytical Electron Microscopy . . . . . . . . . . . . . . . . . . . . . 214

6 References . . . . . . . . . . . . . . . . . . . . . . . . . . . . . . . 217

# 1 Thickness Distribution in Linear PE-Material

Most polymers are damaged chemically when exposed to the electron beam, whereby the exact nature of the damage depends on the chemical consituents of the chain molecules. Furthermore, polymers show only very weak contrast because the atomic numbers of the scattering atoms are generally very low. For this reason microscopists have had to develop a number of techniques in order to overcome these handicaps. These include, amongst others, the use of low temperatures, image intensification techniques and chemical staining. In the latter technique either one component in a mixture of different polymers or the amorphous part of a semi-crystalline polymer is preferentially stained with a heavy element. The latter technique certainly changes this component of the material chemically; however it does not necessarily change the morphology. On the contrary, in some cases a chemical stain can serve to freeze-in the structure and actually stabilise the specimen in the electron beam. In the first section experiments on chemically stained polyethylene will be described.

In order to assess whether electron microscopy on stained polymers is a suitable technique for numerical analysis, detailed comparative studies with Fourier transformed X-ray small angle scattering (SAXS) and Raman longitudinal acoustic mode (LAM) spectroscopy were undertaken. Initial results from comparative measurements were presented previously [1,2]. These have since been extended and a considerable amount of detailed information has emerged [3,4,5], all of which has indicated that the processes controlling crystallization are not only temperature dependent but also, in the case of isothermal crystallization, molecular weight dependent. Consequently it is mandatory to work with fractions in order to avoid overlap of different parameters. The fractions used here were all prepared and thermally characterised in Professor Mandelkern's laboratory, where most of the Raman measurements were also performed. The electron microscopy and X-ray analysis were performed in Mainz. For electron microscopy, thin sections and a staining technique first proposed by Kanig [6a] was used. Contrast is obtained only from lamellae which are parallel to the electron beam. Optimum contrast can be obtained by tilting. Under these conditions the lamellar thicknesses and long spacings can be evaluated directly from the micrographs and plotted in the form of histograms. The following samples were investigated:

a) Rapidly quenched fractions of bulk linear polyethylene with molecular weights ranging from $M = 5 \times 10^3$ to $M = 6 \times 10^6$;

b) Isothermally crystallized fractions of bulk linear polyethylene with molecular weights ranging from $M = 5 \times 10^3$ to $M = 6 \times 10^6$.

a) Fig. 1 shows that the crystallite morphology changes very characteristically with molecular weight for quenched samples. While the crystal thickness does not appear to be significantly affected by molecular weight, the lateral extension of the crystallites is remarkably long at low molecular weights and decreases dramatically as the molecular weight increases.

Furthermore, intralamellar contrast increases. Finally, at very high molecular weights ($M_w = 1.6 \times 10^6$) only very short crystallite segments are observed. The histograms presented in Fig. 2 show that the crystallite thickness distribution is fairly narrow and that the measured values appear to be unaffected by molecular weight. This is clearly indicated in Fig. 3, where the maximum, average, and minimum crystal

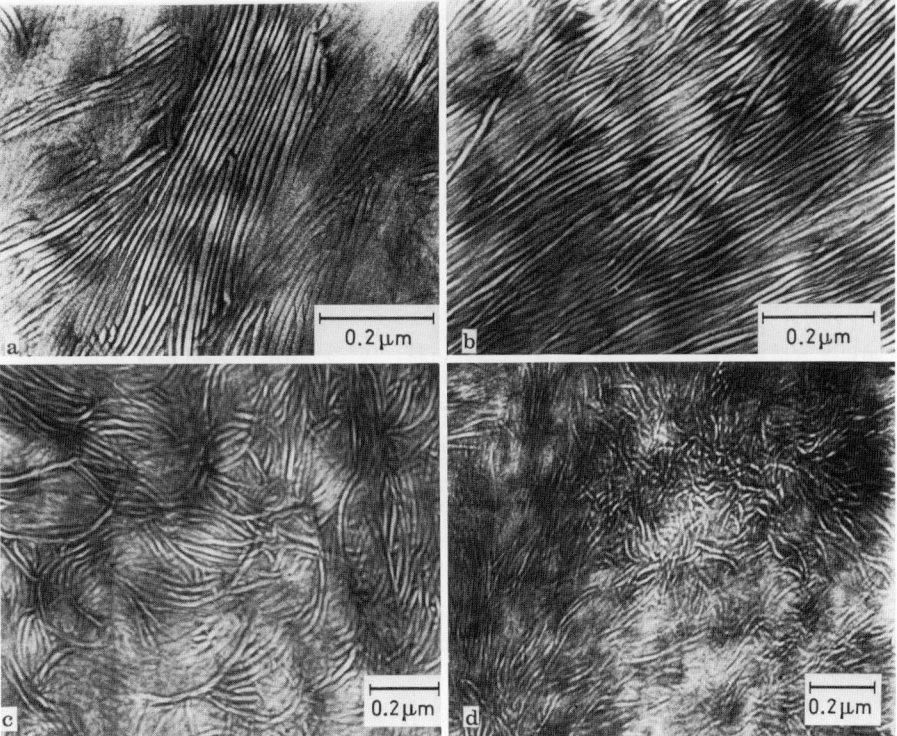

**Fig. 1a–d.** Morphology of quenched polyethylene fractions. **a)** $M = 5 \cdot 6 \times 10^3$; **b)** $M = 1 \cdot 1 \times 10^4$; **c)** $M = 4 \cdot 6 \times 10^4$; **d)** $M = 1 \cdot 89 \times 10^5$

thicknesses are plotted as a function of molecular weight. Except for very low molecular weights, where crystal thickness corresponds to the extended chain length, there is no molecular weight effect and the average value remains constant at about 100 Å. In order to substantiate the numerical values, Raman distributions are shown from the same samples (Fig. 4). The crystal thickness as calculated from Raman measurements also remains constant as a function of molecular weight, but at a slightly larger value. Such a shift in LAM frequency has also been observed by Strobl [6b], who has shown both theoretically and experimentally that the assumption of a complete decoupling of the LAM motion at the lamellar surface is not always justified and is particularly large for thin lamellae with rough surfaces. These electron microscopic results substantiate this conclusion.

While the crystal thickness after quenching is unaffected as molecular weight is increased, this is not true for the long spacings, which increase (Fig. 5). The increasing amorphous thicknesses and decreasing lateral extension of lamellae with increasing molecular weight is an indication that entanglements are increasingly pushed first to the surfaces and then the lateral edges of the crystals. In all cases where narrow thickness distributions are obtained, the agreement between the results obtained from electronmicroscopy, Raman LAM and X-ray small angle scattering, is good.

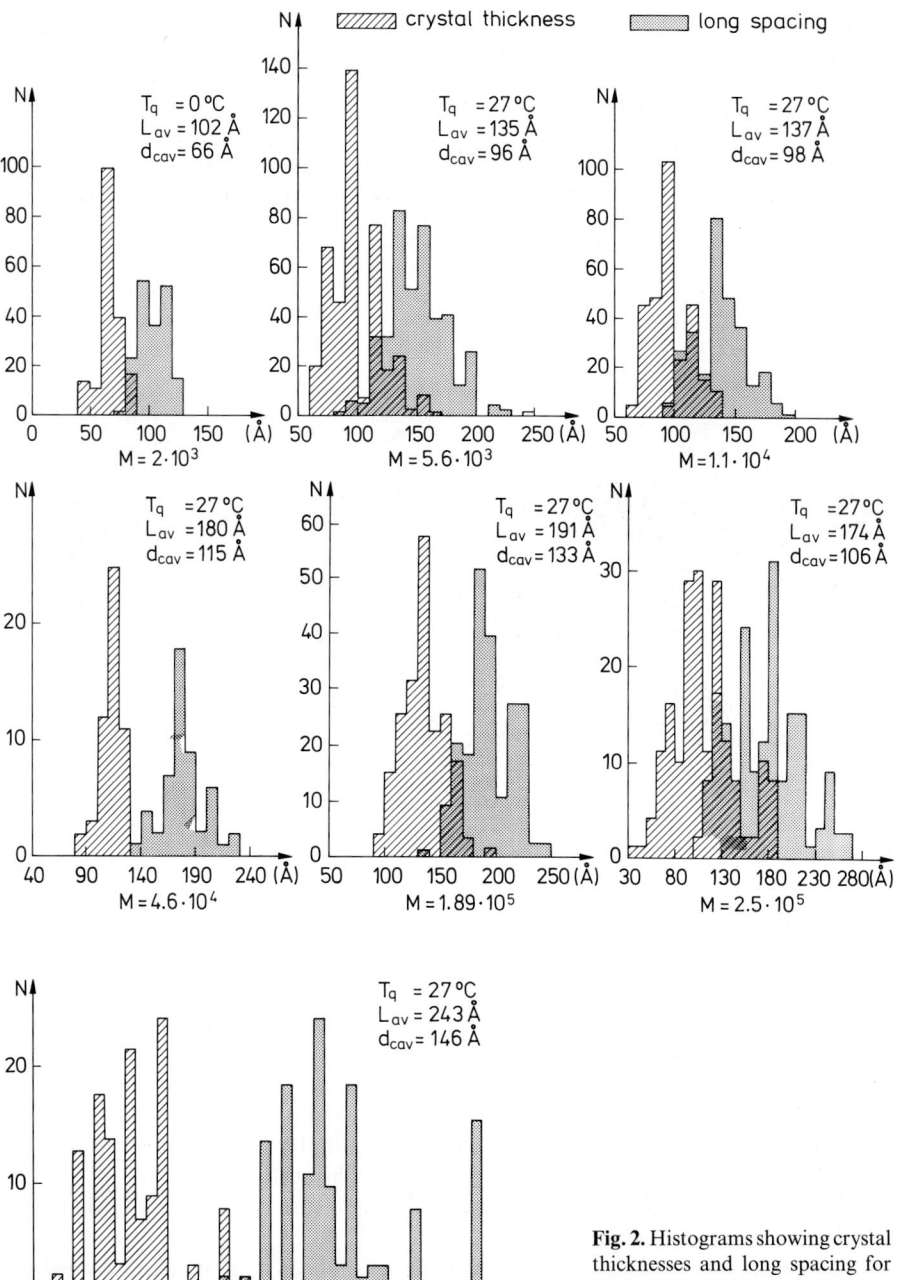

**Fig. 2.** Histograms showing crystal thicknesses and long spacing for different molecular weights

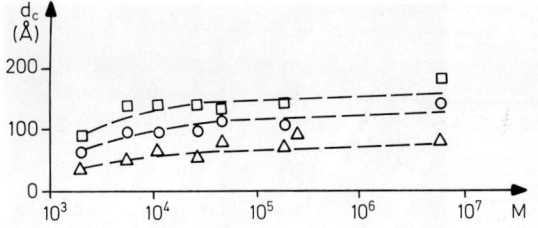

**Fig. 3.** Dependence of crystal thickness on molecular weight for rapid quench; □$d_{c\,max}$, ○$d_{c\,av}$, △$d_{c\,min}$

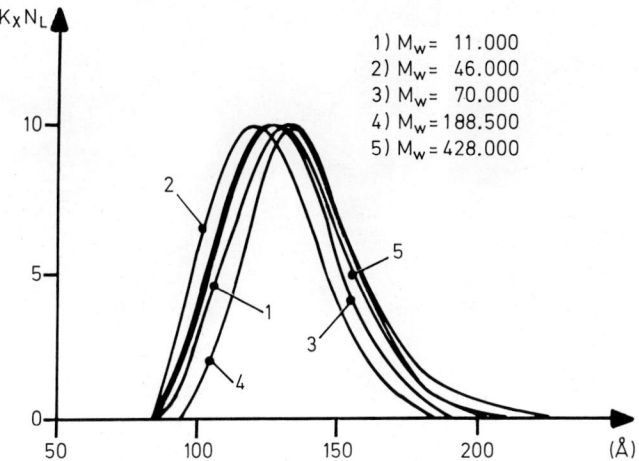

1) $M_w$ = 11.000
2) $M_w$ = 46.000
3) $M_w$ = 70.000
4) $M_w$ = 188.500
5) $M_w$ = 428.000

**Fig. 4.** Distribution of $CH_2$ trans lengths for polyethylene fractions fast quenched into n-pentane

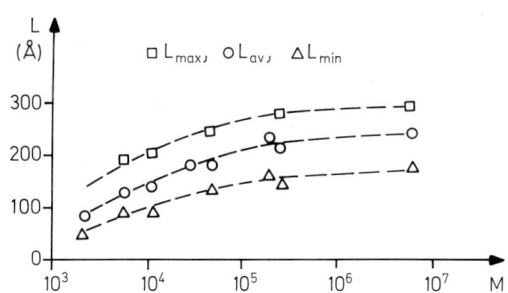

□$L_{max}$, ○$L_{av}$, △$L_{min}$

**Fig. 5.** Dependence of long spacing on molecular weight for rapid quench

b) The situation regarding isothermal crystallization at low undercooling is entirely different. The crystals are considerably thicker (Fig. 6) and even before numerical analysis is given, the complicated nature of the thickness distributions is obvious by mere inspection of the micrographs. The thickest crystals are observed at a molecular weight of $0.7 \times 10^4$. At the same time, the lateral extension of the lamellae decreases with increasing molecular weight while line contrast within the lamellae increases.

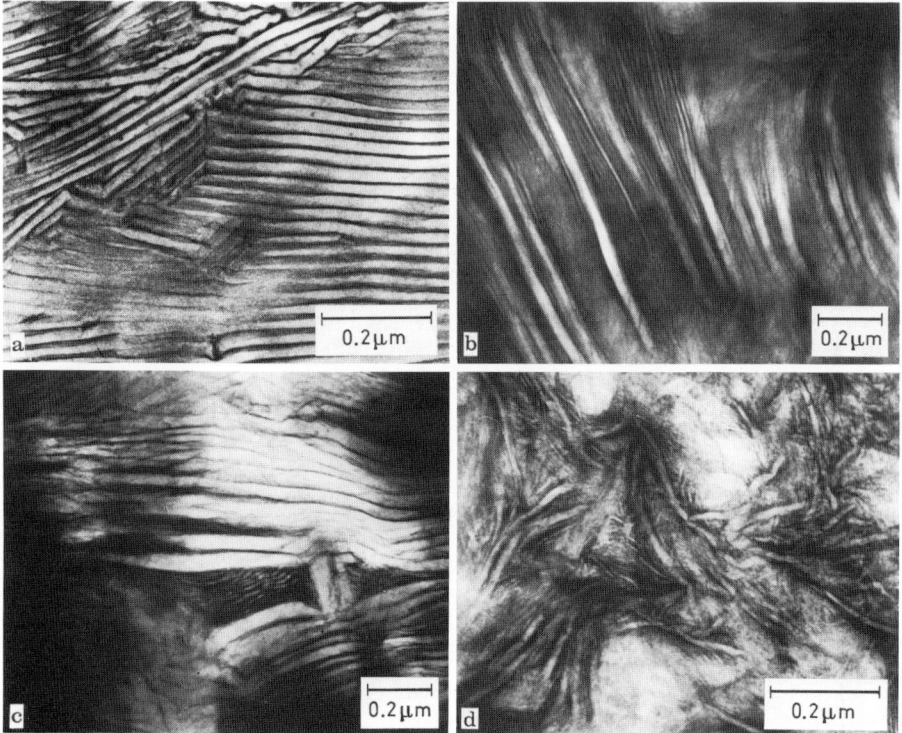

**Fig. 6a–d.** Morphology of isothermally crystallised polyethylene fractions.
**a)** M = 5.6 · 10³, Tc = 127 °C; **b)** M = 8.6 · 10³, Tc = 128.5 °C; **c)** M = 2.7 × 10⁴, Tc = 130 °C; **d)** M = 1.6 × 10⁶, Tc = 130 °C

The histograms of Fig. 7 indicate that the distributions are extremely broad and complex.

After removal of the contribution from crystals formed on quenching subsequent to the isothermal crystallization (left of the dotted line in the histograms) the distributions are still very broad, giving a factor of 4—5 between the thinnest and thickest crystals. The manner of representation chosen in Fig. 8 shows that this difference reaches a maximum at a molecular weight of about $2.7 \times 10^4$, where crystals over 1000 Å have been achieved. The explanation for these very broad distributions, which we have suggested previously [3] and have since substantiated by other measurements [4] is isothermal thickening. This is fully in accord with the conclusions reached by other authors [7] and in fact even the numerical values for the thickness of the thinnest crystals ($d_{cmin}$) are in agreement with the values of the "original crystallite" size found by these authors, although an entirely different technique was used. Maximum thickening is achieved at intermediate molecular weights, beyond which this process is prevented by entanglements at the lamellar surfaces.

Inspection of the micrographs immediately indicates that the numerical results obtained by each of the measuring techniques employed will differ. The thick isolated lamellare are not detected in the small angle X-ray scattering peak, which averages

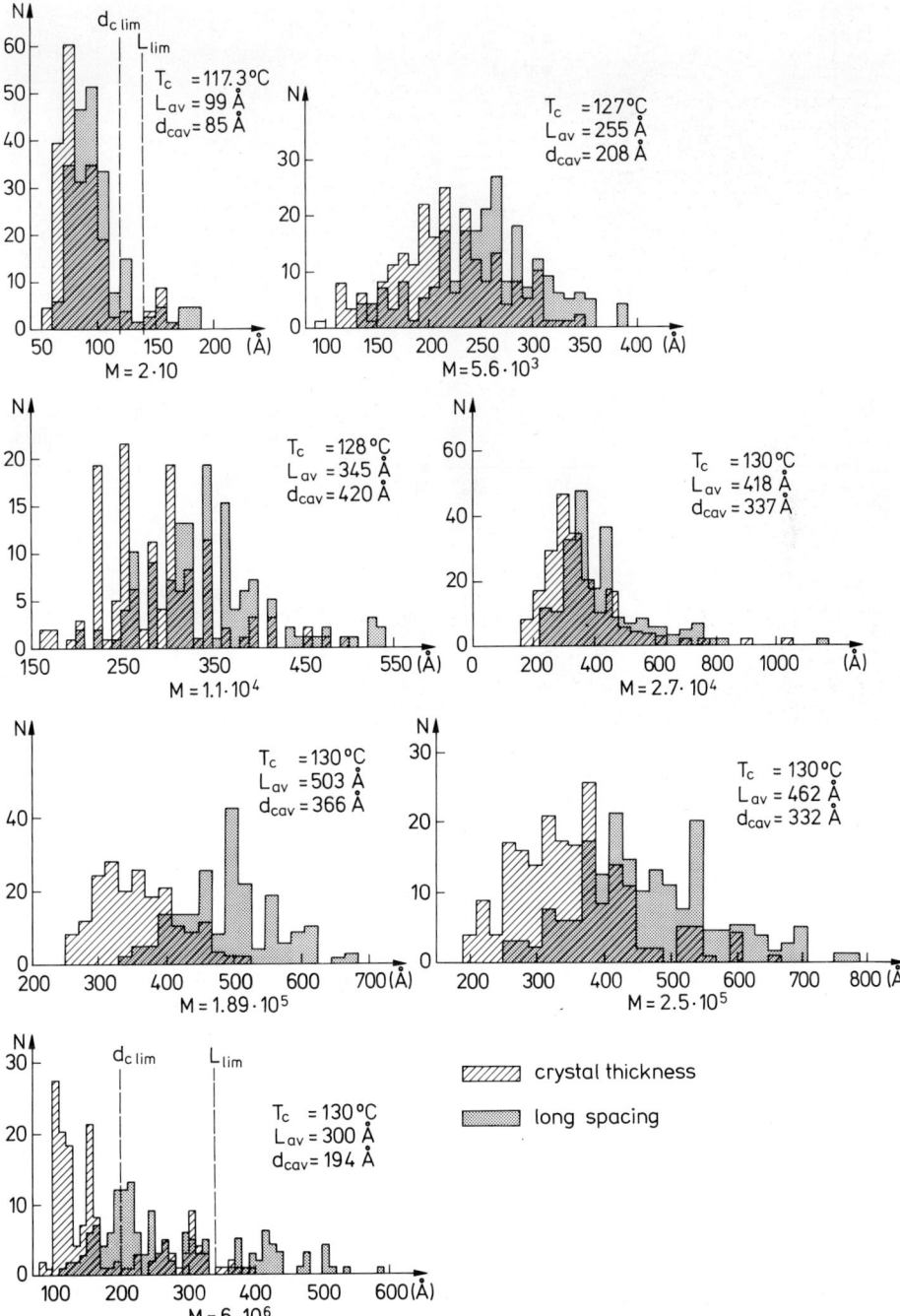

**Fig. 7.** Histograms showing crystal thicknesses and long spacings for different molecular weights

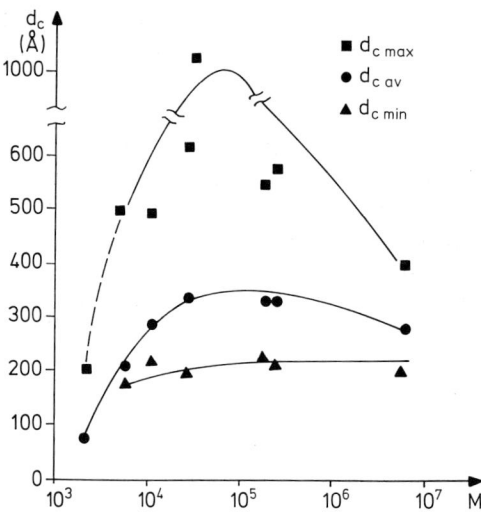

**Fig. 8.** Dependence of crystal thickness on molecular weight for isothermal crystallisation

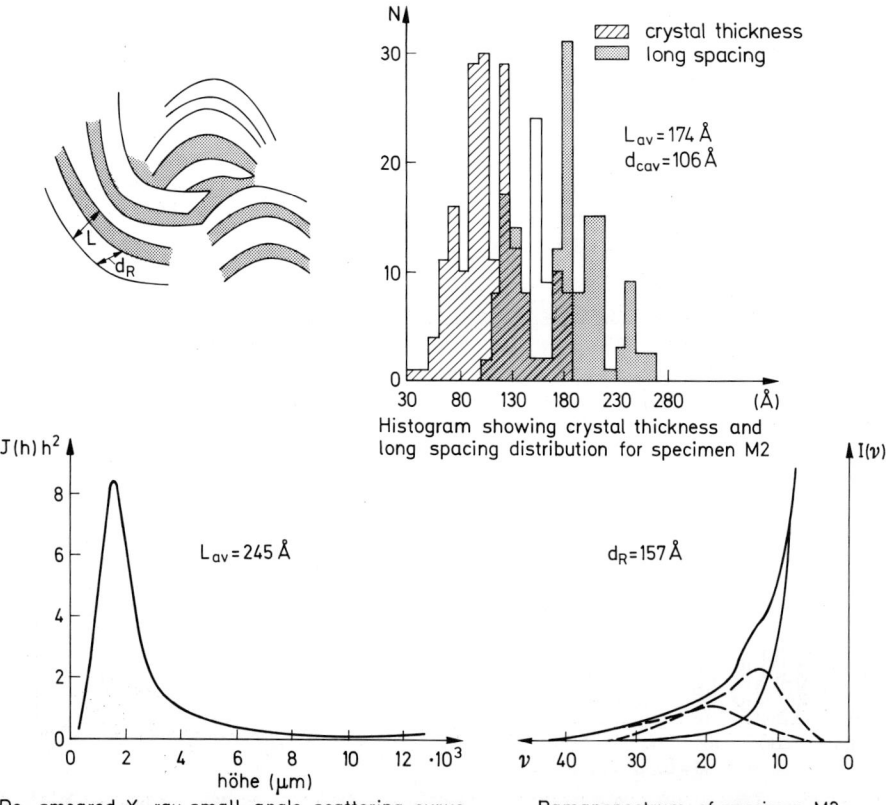

**Fig. 9.** Comparison between Raman, EM and SAXS measurements on curved crystals

only those values obtained from lamellar stacks. In contrast to this, the Raman signal from such crystals will be strong, while the very broad distribution of thin crystals is difficult to separate from the background. Also, the shift in peak maximum can be considerable for broad distributions when correcting from weight average for comparison purposes. Furthermore, by electron microscopy, microscopic areas are analysed, which may not be representative of the bulk. Finally, large, thick crystals are only rarely in the correct orientation for contrast, so that they are underrepresented in number, while very small thin crystals tend to overlap, also causing reduced contrast. Consequently for isothermally crystallized samples having broad distributions it is not surprising to observe large differences in numerical values obtained by different techniques. This is demonstrated for two typical morphologies and distributions in Figs. 9 and 10. It is also reflected in the comparison between the crystallinity obtained from density measurements and that obtained from the values of $\dfrac{d_{cav}}{L_{av}}$ obtained from

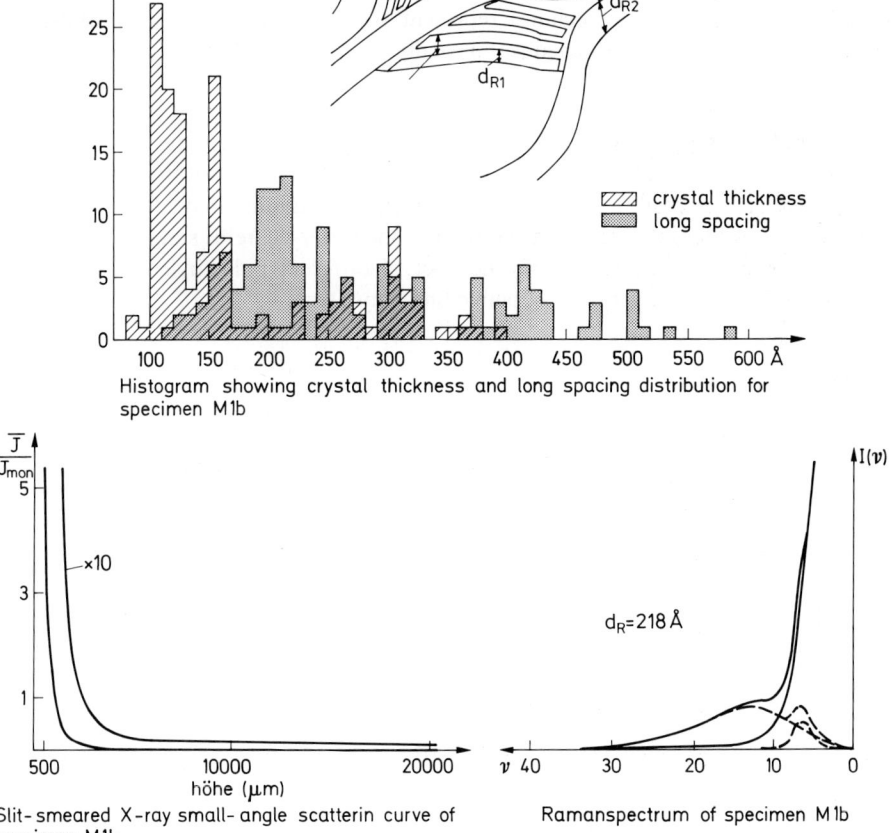

Fig. 10. Comparison between Raman, EM and SAXS for extremely broad distribution

electron microscopy (Fig. 11). While the values obtained for the quenched samples are in good agreement, those relating to the isothermally crystallized material only agree approximately, while however, giving the correct trend. This is not surprising in view of the fact that it is meaningless to assess an average value of a broad non-symmetrical distribution which is known to be incorrectly weighted.

In Fig. 12 it is demonstrated that the dramatic increase in thickness of the thickest crystals begins at 125 °C for all linear polyethylene fractions while $d_{cmin}$ does not increase to the same extent. Thickening may not affect the whole crystal but only those parts of it which are geometrically favourably orientated with respect to the surrounding crystals (Fig. 13). This thickening process has been assessed numerically in Fig. 14 as a function of time. Although we are aware that the distribution is not correctly weighted, it is clear that the proportion of thin crystals of about 100 Å decreases, while the maximum value relating to the thicker crystals has moved from 200 Å to 300 Å.

It is significant not only to note the numerical trends of the thickest crystals $d_{c\,max}$ but also those of the thinnest crystals $d_{c\,min}$ as a function of undercooling (Fig. 15). In this case a linear relationship is obtained. This finding substantiates the conclusion that the thinnest crystals observed in the micrograph (after subtraction of the known distribution formed on quenching) represent the original nuclear size, whilst the remaining crystals have thickened isothermally.

## 2 Thickness Distribution in Co-Polymers

In the case of branched polyethylene the danger that staining may affect the crystal surfaces is considerably increased. However, under very careful staining conditions the experimental results for a large number of different polyethylene co-polymers have been found to be very reproducible and consistent.

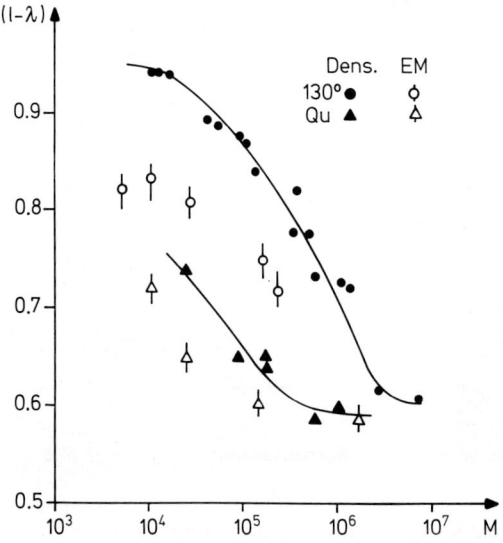

**Fig. 11.** Dependence of crystallinity on molecular weight

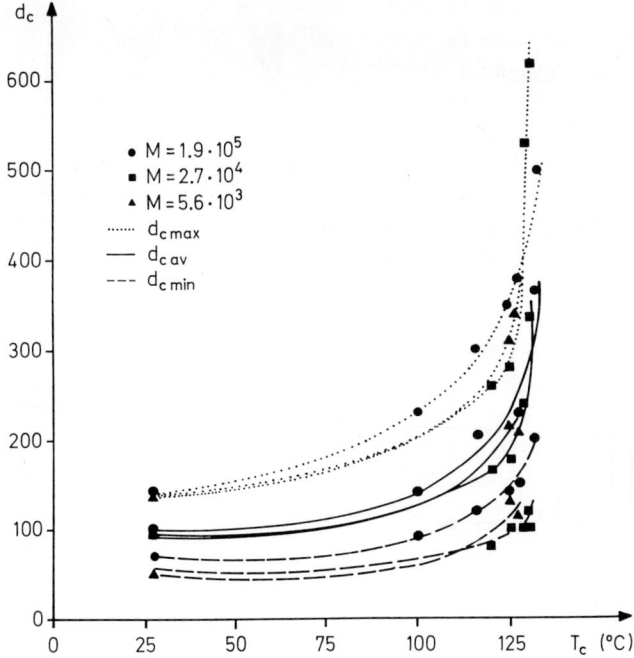

Fig. 12. Dependence of crystal thickness on crystallisation temperature

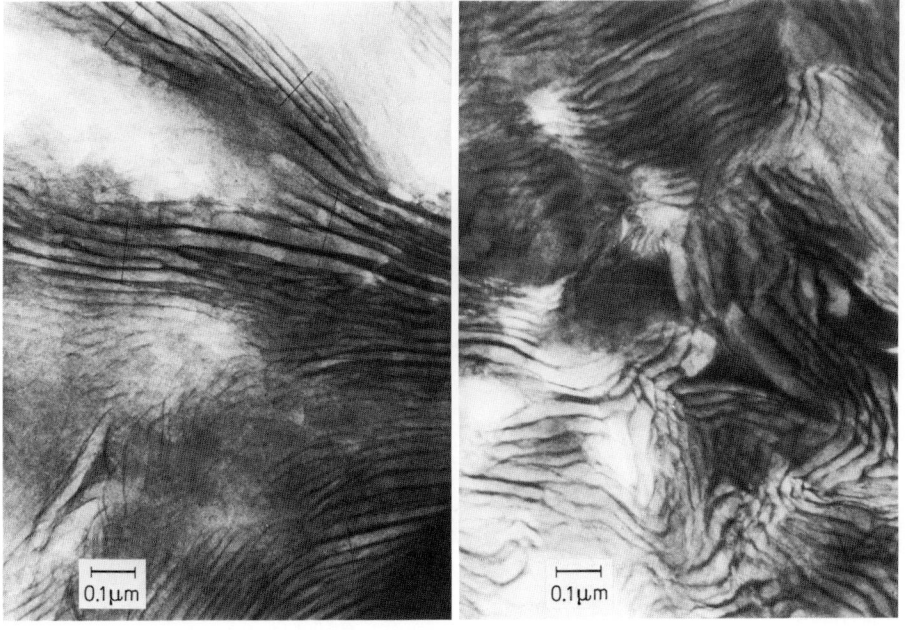

Fig. 13. Micrographs indicating crystal thickness distribution at high crystallisation temperatures

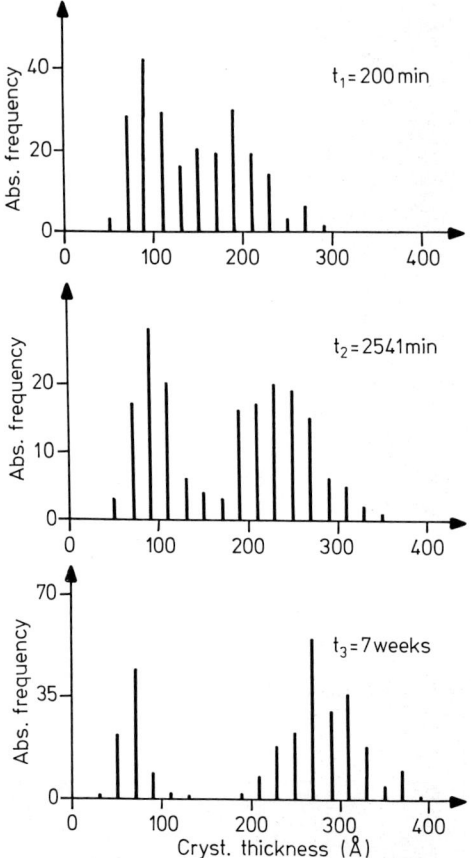

**Fig. 14.** Dependence of crystal thickness as a function of crystallisation time at constant temperature for linear polyethylene fraction

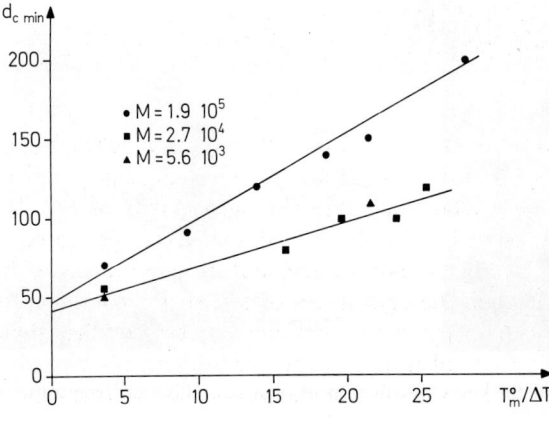

**Fig. 15.** Dependence of minimum crystal thickness on undercooling

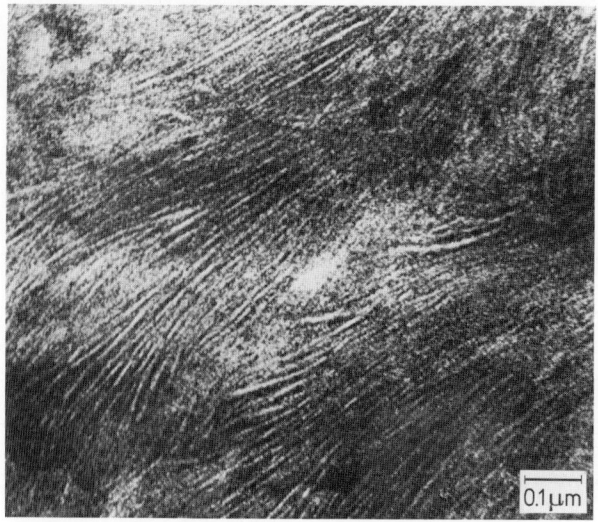

Fig. 16. Branched Polyethylene $T_c = 100\,°C$, $T_a = 25\,°C$

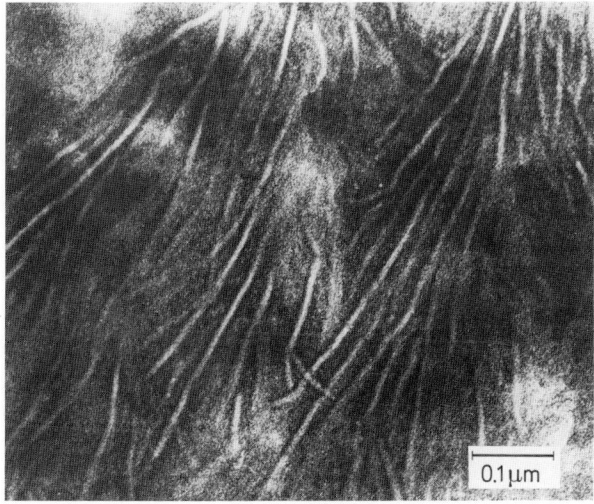

Fig. 17. Branched Polyethylene $T_c = 100\,°C$, $T_a = 100\,°C$

In contrast to linear material, it is not the crystal thickness distribution which is broad, but the amorphous thickness distribution. Furthermore, it is this quantity which changes on annealing and not the crystal thickness. The micrographs of Fig. 16 (non-annealed) and Fig. 17 (annealed at 100 °C) show the dramatic change in amorphous thickness. The histograms in Fig. 18 indicate the shift in long spacing with increasing annealing temperature while the crystal thickness remains constant. It should be noted that the long spacing distribution is considerably broader than that of the crystal thickness, though not as broad as that observed in isothermally crystallized linear material. Again it is the value of the smallest and largest long spacing which

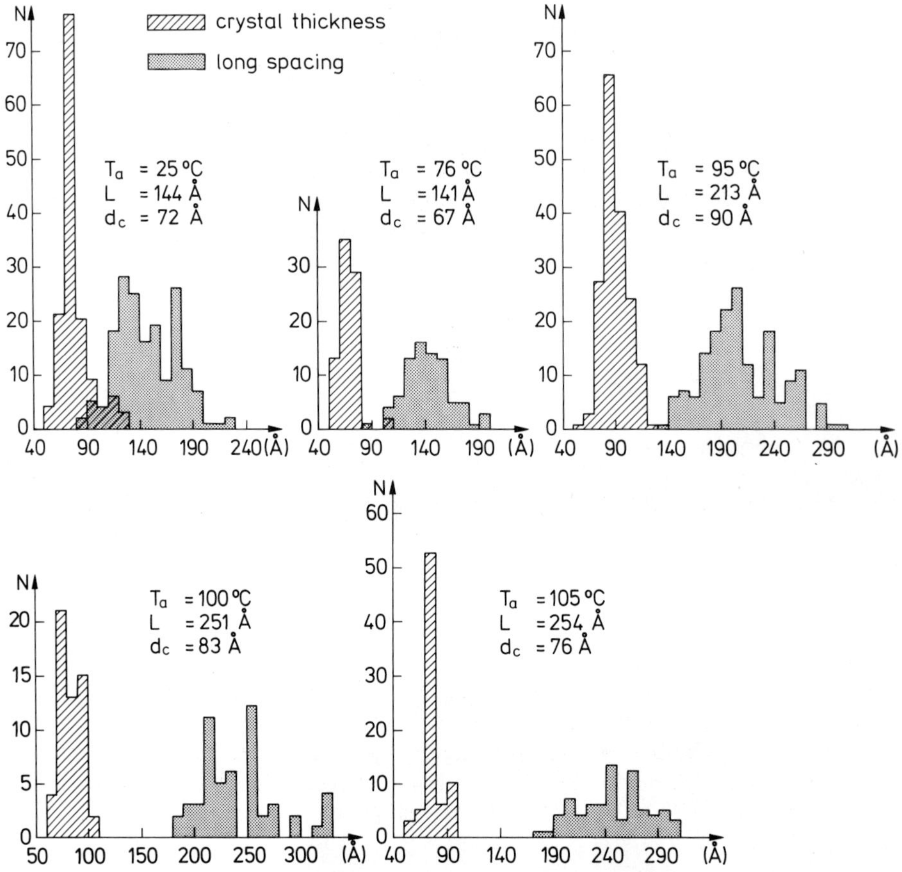

**Fig. 18.** Branched Polyethylene. Long spacing (L) and crystal thickness ($d_c$) for various temperatures. Histograms obtained from electron micrographs

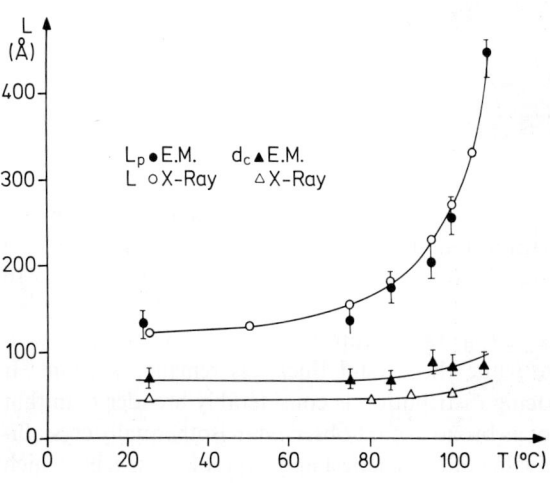

**Fig. 19.** Comparison of long spacings and crystal thicknesses obtained by E.M. and SAXS

is significant and offers a strong argument in favour of the "insertion" model für crystallization and melting proposed by us previously [2]. The agreement between electron microscopy and Fourier transformed X-ray scattering results for the average long spacings was found to be excellent (Fig. 19). The consistently slightly higher values for crystal thickness can easily be explained by the assumption that side groups are accumulated at the crystal surfaces, a very likely event in view of the fact that the crystal thickness in branched material is found to correspond to the average distance between branches, and the staining medium cannot penetrate these regions. It is these side groups which then prevent the thickening mechanism, which is operative in linear material, from taking place.

## 3 Evaluation of Molecular Tilt Angle

It is observed in all isothermally crystallized samples (except at the highest molecular weight) and in some cases for low molecular weights, that crystals are not curved but "roof shaped" [3]. As shown in Fig. 20, the "roofs" are often formed in a cooperative manner. Analysis of the measured "apex" angles (Fig. 21) shows that certain angles appear in large proportion, namely those centering around 90, 110, 135, 125, 135 and 145° and that the percentage of each depends on the crystallization temperature. Considering the relationship between the orthohombic polyethylene unit cell and the observed angles, the surface planes indicated in Fig. 22 can be calculated. Remembering that a "roof" is only visible if both planes are simulatenously parallel to the electron beam (i.e. the roof gable is perpendicular to the thin section) then the experimentally observed apex angles can be explained by a combination of surface planes as indicated in Fig. 23. It turns out that all the observed experimental values can be explained by a combination of (h0l) planes, so that the roofs are observed

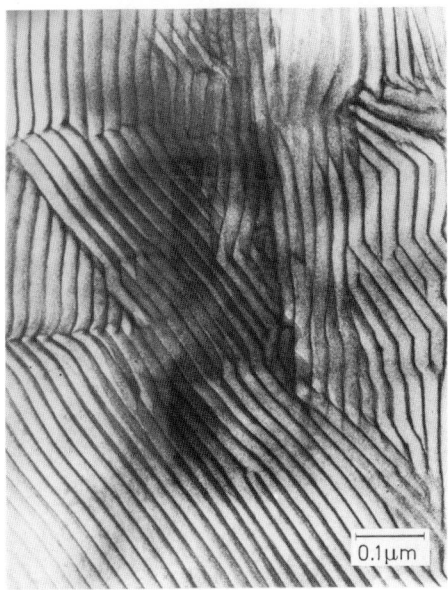

**Fig. 20.** Detail in low molecular weight ($M_w = 5.6 \times 10^3$) polyethylene fraction ($T_c = 125$ °C)

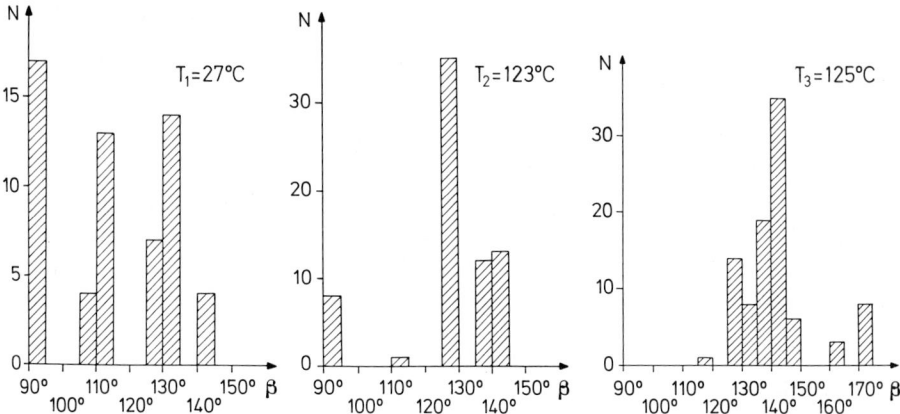

**Fig. 21.** Analysis of apex angles at various temperatures

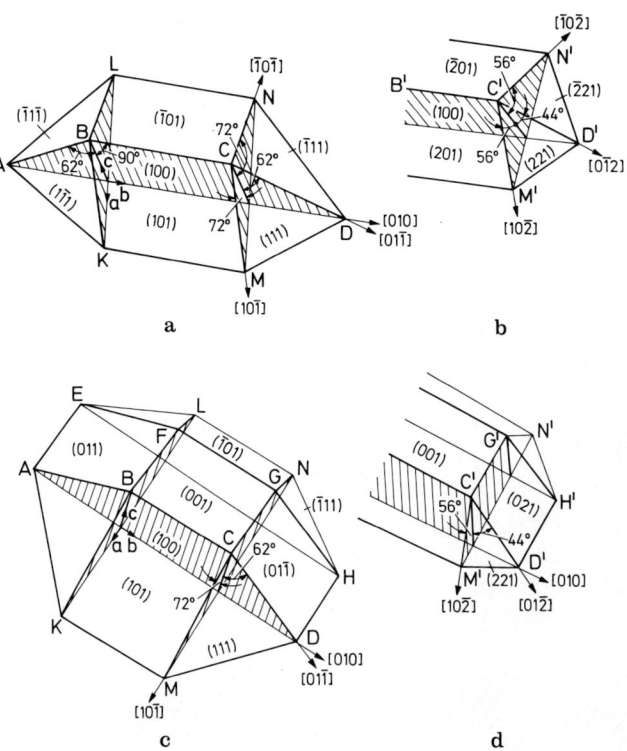

**Fig. 22a–d.** Crystallographic relationships in polyethylene relevant to observed structures in electron micrographs

Use of Transmission Electron Microscopy 211

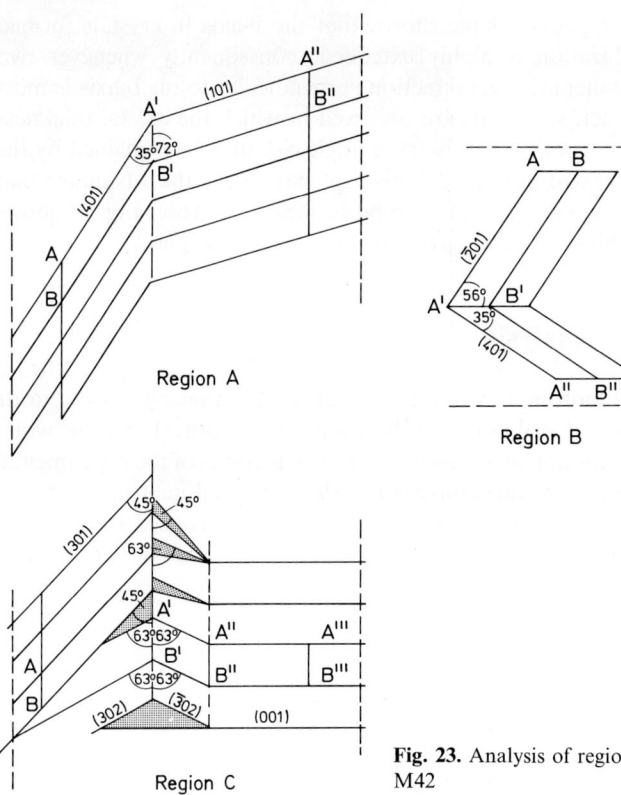

Fig. 23. Analysis of regions A, B and C from specimen M42

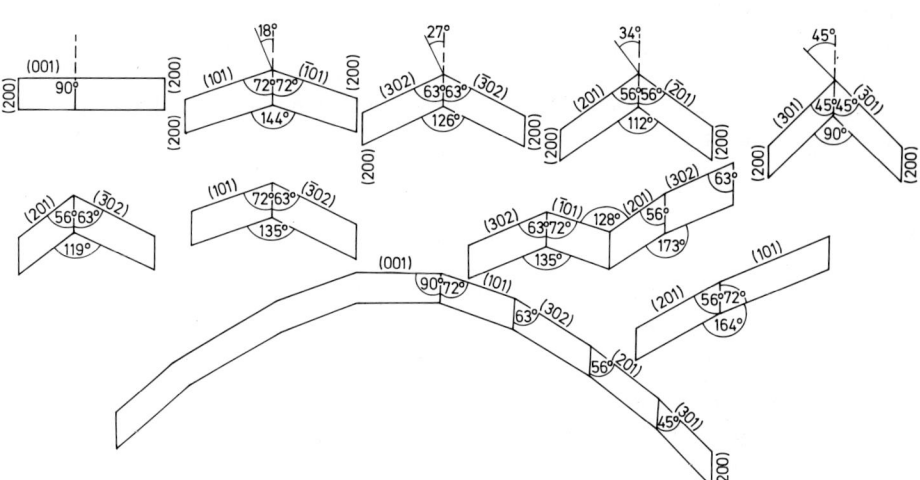

Fig. 24. Molecular tilts calculated from observed apex angles for specimen M43 indicating transition from "roof shaped" to curved morphology

along the b-axis. Surface replicas [8] have shown that the b-axis in crystals formed during isothermal crystallization is highly extended; consequently whenever two planes are observed simultaneously, the direction perpendicular to the b-axis is most likely to be observed. In fact, situations are observed in which the crystal thickness varies on both sides of the "roof". The observed angle can then be explained by the combination of planes indicated in Fig. 23. This explanation has the advantage that the molecular stem length remains constant on both sides of the roof. Fig. 24 shows most of the possible combinations giving rise to the observed angles.

## 4 Radial Distribution Analysis

Diffraction patterns from amorphous materials consist of a few vague halos. Accurate measurement of the intensity distributions of these halos in reciprocal space provides information about the distribution of atoms in real space. Reviews of the experimental methods and data handling techniques for X-ray-, electron-, and neutron-scattering is given in the literature [9, 10, 11]. The mathematical tool which is used to relate the scattering function $S(q, \omega)$, with the correlation function $G(r, t)$ is a Fourier transformation:

$$S(\vec{q}, \omega) = \frac{1}{2\pi} \int e^{i(\vec{q}\cdot\vec{r} - \omega t)} [G(\vec{r}, t) - \varrho] \, d\vec{r} \, dt$$

The intensity is given by

$$I(q, t) = \int_{-\infty}^{+\infty} S(q, \omega) \, e^{i\omega t} \, d\omega$$

In the above relationships $q = \frac{2\pi}{\lambda} \sin \theta$ is the wave vector of the scattered radiation, is the wavelength and $\omega$ is the frequency shift.

In the case of electron diffraction, we consider the coherent, elastically scattered electrons only, so that the time-independent, static correlation function $G(r, 0)$ is obtained in terms of the number density $\varrho(r)$ of atoms on the surface of a sphere a certain distance r from an atom at the origin:

$$I(q) = Nf(q)^2 \left\{ 1 + \int_0^\infty \frac{\sin qr}{qr} [\varrho - \langle\varrho\rangle] 4\pi r^2 \, dr \right\}$$

$$G(r, 0) = 4\pi r^2 (\varrho(r) - \langle\varrho\rangle) = \frac{2r}{\pi} \int_0^\infty q \left[ \frac{I}{Nf^2} - 1 \right] \sin qr \, dq$$

Such investigations using electron diffraction (Fig. 25) have been performed on polyethylene melts (12) and the appearance of sharp maxima in the correlation function indicating the intra-molecular, as well as a broad maximum indicating a distribution of inter-molecular, distances in the intermediate distance range could be demonstrated (Fig. 26). Due to the development of sophisticated neutron scattering techniques, offering a much larger range of experimentally accessible scattering

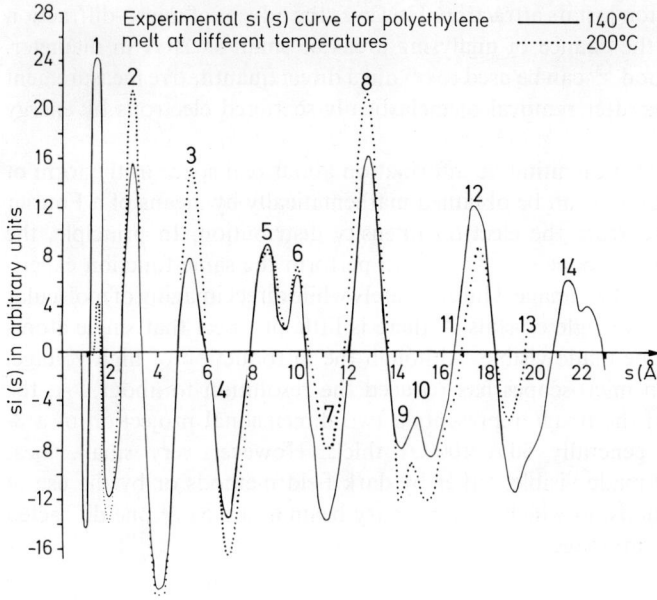

**Fig. 25.** Experimental si(s) curve for polyethylene melt at different temperatures

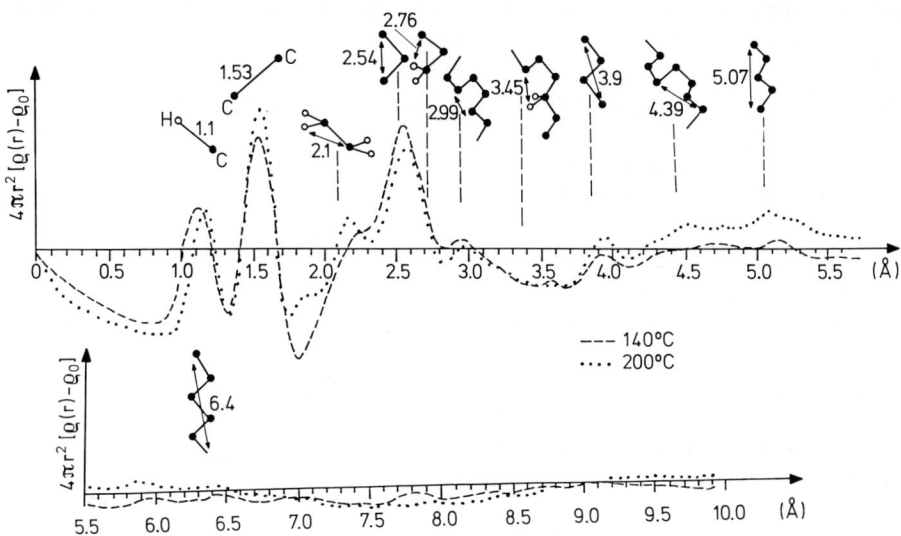

**Fig. 26.** Experimental radial distribution curves

angles, electron diffraction lost its attraction. However the advent of micro-diffraction techniques now offers the chance of analysing areas as small as 20 Å in diameter. A special scanning method [13] can be used to obtain a direct quantitative measurement of diffraction intensities after removal of inelastically scattered electrons by energy filtering.

We have shown above that quantitative information about real space in the form of a radial distribution function can be obtained mathematically by means of a Fourier transformation obtained from the electron intensity distribution. In principle, the imaging lenses of the microscope can be used to perform the same function experimentally in the formation of an image. Unfortunately while direct imaging of molecular chains can be achieved in single crystals [14] there is little prospect that single atoms in a random array can be made visible. Although the introduction of high voltage, high resolution electron microscopes has reduced the resolution to about 2 Å, the main problem remains; the image represents a two dimensional projection of a 3-dimensional structure generally 50 Å–1000 Å thick. However very small, local ordered regions can be made visible, either by dark field methods or by the use of interference fringe methods, in which both primary beam and part of one diffracted beam are used to form an image [15].

## 5 Analytical Electron Microscopy

A paper on new developments in the characterisation of polymers in the solid state must include a discussion of the possibilities offered by scanning electron transmission and the ancillary detection devices EDX (energy dispersive analysis of X-rays) and EELS (electron energy loss spectroscopy).

In scanning transmission the electron beam from a small, bright source is demagnified by the condensor lenses and converges to form a beam of very small cross-section on the specimen, beyond which it diverges again to form a circular spot in the detector plane. The diffraction pattern formed is a convergent beam electron diffraction pattern. To obtain an image, part of this pattern is detected by a surface barrier detector and the resulting image signal is displayed on a cathode ray tube as the incident beam is scanned across the specimen. Both bright and dark field imaging techniques can be used, depending on whether the detector aperture selects the central portion of the convergent beam diffraction pattern or the portion scattered at an angle $2\theta$. Modern instruments offer micro-diffraction facilities, making it possible, in principle, to obtain diffraction patterns from areas as small as 20 Å in diameter. This is achieved if the incident beam in a STEM instrument is not scanned over the specimen but held stationary. The beam damage incurred to the specimen is a factor limiting the potential of the technique.

Apart from the elastically scattered electrons which are used to form an image, a number of other signals are generated by the interaction of an energetic electron beam with a specimen (Fig. 27). While most of the primary electrons are used to form an image, a certain fraction excites the atoms of the specimen, which then emit X-rays when they return to their ground state. The characteristic X-rays generated in the sample can be used to obtain elemental analysis of the irradiated region. In scanning electron microscopes, energy dispersive spectrometers are used for this purpose,

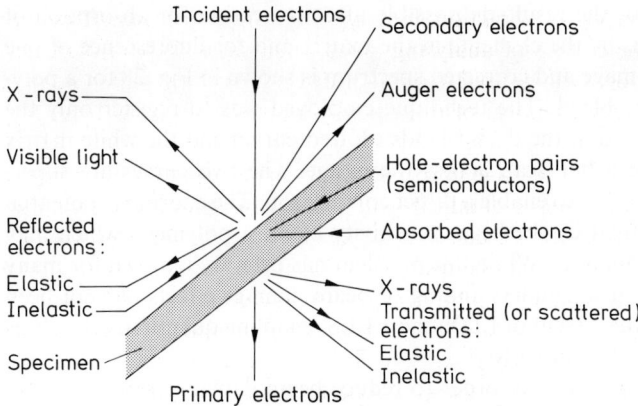

**Fig. 27.** Signals generated through the interaction of an energetic electron beam and a thin specimen

because they are better suited to the low X-ray fluxes encountered in STEM operation than a wavelength dispersive spectrometer. They have the disadvantage that elemental analysis is only possible down to atomic numbers greater than nine. Recently a windowless detector has been developed for STEM operation, in which case elements down to Boron can be detected. Various types of analysis can be chosen:

a) X-ray distribution images, where the X-ray analyser is set for a particular element of interest. Whenever an X-ray is detected of that energy, a dot is brightened at the corresponding point of the display tube.

b) X-ray line scans, in which case the beam is scanned along a single line and the intensity from a single element is plotted directly on the STEM display.

c) Point analysis, where the electron beam is positioned on a feature of interest and remains stationary while the analysis is obtained.

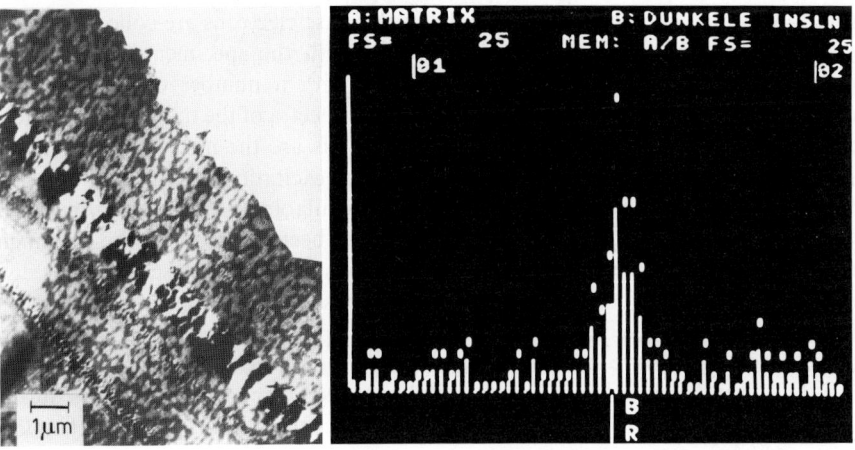

**Fig. 28.** Polymer Blend, Polystyrene/Polybromostyrene 47:53 with energy dispersive X-ray spectrum

Quantitative analysis of the results is possible after correction for absorption of X-rays from each element by the elements in the matrix and for fluorescence of one element by another. An image and corrected spectrum is shown in Fig. 28 for a polystyrene/polybromostyrene blend. The technique employed was to register only the bromine which was detected in the dark islands (dotted curve) and the white matrix (continuous lines) during a line scan across the image. The two curves are superimposed on top of one another to enable a direct comparison. The enormous potential of this technique is inhibited by the beam sensitivity of most polymers, which may lead to mass loss during analysis. While this problem has been well known for many years [16, 17] real progress in the understanding of beam damage effects has not been made until recently with the advent of EDX and EELS, enabling quantitative analysis of mass loss to be undertaken directly [18].

Factors which are important [19] in order to reduce beam damage are:

1) low brightness electron gun with a good gun bias control;
2) an electron beam shutter above the specimen;
3) bright/dark field STEM deflector close to the projector lens apperture;
4) double scanning coil with a possibility to de-scan the beam in the objective back focal plane;
5) full control of condensor and imaging lenses;
6) efficient detection and recording system;
7) pre-selected modes of operation with a quick change-over;
8) an annular detector and signal mixing device;
9) high accelerating voltage;
10) cooling stage for temperature below 10 °K.

The last requirement is still a subject of debate. While some experiments seemed to indicate that mass loss could be considerably reduced at low temperatures [20, 21], other workers have found that very little life time is gained below liquid nitrogen temperature, which can easily be achieved using a standard cooling holder.

A technique which appeared to offer very promising applications in polymer science is that of electron energy loss spectroscopy (EELS) because it enables elements of very low atomic number to be analysed. In this case electrons are collected which have suffered an energy loss during interaction with the specimen. Considerable experience with this technique has been gained with a number of biomolecular compounds [22]. The regions in which the energy loss spectra of the transmitted 25 KeV electrons are particularly useful for chemical analysis are the inner shell excitation region (energy loss ~ 50–400 eV) and the valence shell excitation (energy loss < 50 eV). Unfortunately, in the spectrometers presently available, a sequential detection technique is used, so that considerable damage has been incurred by the specimen before the element of interest can be analysed. Consequently the use of energy dispersive X-ray analysis in conjunction with windowless detection appears to be a more fruitful technique for detecting light elements at the present time. Detailed measurements of decay rate have been performed on PVC, PVDC, PTFE [23]. This work shows that an understanding of the chemical processes occuring in polymers during beam exposure can also be used in a positive sense to promote specific chemical reactions which can be used to obtain image contrast in specimens which would otherwise appear homogeneous.

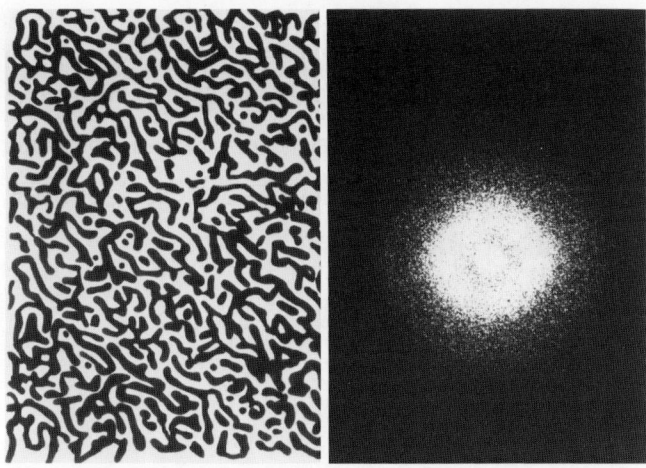

**Fig. 29.** Polystyrene/Polybromostyrene composite with light scattering pattern

*Acknowledgements*: The X-ray spectrum shown in Fig. 28 was obtained during a series of tests undertaken together with Dr. Hagemann at the Philips laboratory in Eindhoven. The particular specimen shown is a polystyrene/polybromostyrene blend obtained from Professor Strobl, University of Mainz. The Raman and density measurements were performed in the laboratory of Professor Mandelkern in Tallahassee.

# 6 References

1. Voigt-Martin, I. G., Fischer, E. W., Hagedorn, W., Hendra, P., Mandelkern, L., Mehler, K.: paper presented at IUPAC, Mainz, p. 1250 (1979)
2. Strobl, G., Schneider, M., Voigt-Martin, I. G.: J. Polym. Sci. (Phys.) *18*, 1368 (1980)
3. Voigt-Martin, I. G., Mandelkern, L.: J. Polym. Sci. (Phys.) *19*, 1769 (1982)
4. Stack, G. M., Mandelkern, L., Voigt-Martin, I. G.: Macromolecules in press (1984)
5. Voigt-Martin, I. G., Mandelkern, L.: J. Polym. Sci. in press (1984)
6a. Kanig, G.: Prog. Coll. Polym. Sci. *57*, 176 (1975)
6b. Strobl, G.: Proceedings of the IUPAC Boston 1982
7. Chivers, R. A., Barham, P. J., Martinez-Salazar, P. J., Keller, A.: J. Polym. Sci. (Phys.) *20*, 1717 (1982)
8. Bassett, D. C., Hodge, A. M.: Proc. Roy. Soc. *A377*, 25 (1981)
9. Wagner, C.: J. Non-Crystalline Solids *31*, 1 (1978)
10. Wright, A. C.: "Advances in Structure Research by Diffraction Methods", Ed. Hoppe, W., Mason, R., Pergamon Press 1974
11. Voigt-Martin, I. G., Wendorff, J.: "Amorphous Polymers" in Mark Encyclopaedia of Science and Technology, revised version 1984
12. Voigt-Martin, I. G., Mijlhoff, F. C.: J. Appl. Phys. *47*, 3942 (1976)
13. Cowley, J. M., Spence, J. C. H.: Ultramicroscopy *3*, 433 (1979)
14. Tsuji, M., Isoda, S., Ohara, M., Kawaguchi, A., Katayama, K.: Polymer *23*, 1568 (1982)
15. Rudee, M. L., Howie, A.: Phil. Mag. *25*, 1001 (1972)
16. Voigt-Martin, I. G., Thesis, Ph. D.: University of London, 1970

17. Grubb, D. T.: J. Mater. Sci. *9*, 1715 (1974)
18. Vesely, D.: EMAG August 1983, Guildford
19. Vesely, D.: paper presented at the EMAG meeting 1981, London
20. Müller, K.-H., Stemmer, A., Zemlin, F., Hermann, K. H.: Tagung für Elektronenmikroskopie, Innsbruck 1981
21. Dietrich, I., Lefranc, G., Müller, K.-H., Stemmer, A.: 7th Eur. Cong. on Electron Microscopy, The Hague *1*, 84 (1980)
22. Crewe, A. V., Isaacson, M. S., Zeitler, E.: "Advances in Structure Research", Vol. 7, Ed. Hoppe, W., Mason, R.
23. Vesely, D., Lindberg, H.: EMAG, August 1983, London


H. H. Kausch (Editor)
Received March 2, 1984

# Author Index Volumes 1–67

*Allegra, G.* and *Bassi, I. W.:* Isomorphism in Synthetic Macromolecular Systems. Vol. 6, pp. 549–574.
*Andrews, E. H.:* Molecular Fracture in Polymers. Vol. 27, pp. 1–66.
*Anufrieva, E. V.* and *Gotlib, Yu. Ya.:* Investigation of Polymers in Solution by Polarized Luminescence. Vol. 40, pp. 1–68.
*Apicella, A., Nicolais, L.* and *de Cataldis, C.:* Characterization of the Morphological Fine Structure of Commercial Thermosetting Resins Through Hygrothermal Experiments. Vol. 66, pp. 189–208.
*Argon, A. S., Cohen, R. E., Gebizlioglu, O. S.* and *Schwier, C.:* Crazing in Block Copolymers and Blends. Vol. 52/53, pp. 275–334
*Arridge, R. C.* and *Barham, P. J.:* Polymer Elasticity. Discrete and Continuum Models. Vol. 46, pp. 67–117.
*Ayrey, G.:* The Use of Isotopes in Polymer Analysis. Vol. 6, pp. 128–148.

*Bässler, H.:* Photopolymerization of Diacetylenes. Vol. 63, pp. 1–48.
*Baldwin, R. L.:* Sedimentation of High Polymers. Vol. 1, pp. 451–511.
*Balta-Calleja, F. J.:* Microhardness Relating to Crystalline Polymers. Vol. 66, pp. 117–148.
*Basedow, A. M.* and *Ebert, K.:* Ultrasonic Degradation of Polymers in Solution. Vol. 22, pp. 83–148.
*Batz, H.-G.:* Polymeric Drugs. Vol. 23, pp. 25–53.
*Bekturov, E. A.* and *Bimendina, L. A.:* Interpolymer Complexes. Vol. 41, pp. 99–147.
*Bergsma, F.* and *Kruissink, Ch. A.:* Ion-Exchange Membranes. Vol. 2, pp. 307–362.
*Berlin, Al. Al., Volfson, S. A.,* and *Enikolopian, N. S.:* Kinetics of Polymerization Processes. Vol. 38, pp. 89–140.
*Berry, G. C.* and *Fox, T. G.:* The Viscosity of Polymers and Their Concentrated Solutions. Vol. 5, pp. 261–357.
*Bevington, J. C.:* Isotopic Methods in Polymer Chemistry. Vol. 2, pp. 1–17.
*Bhuiyan, A. L.:* Some Problems Encountered with Degradation Mechanisms of Addition Polymers. Vol. 47, pp. 1–65.
*Bird, R. B., Warner, Jr., H. R.,* and *Evans, D. C.:* Kinetik Theory and Rheology of Dumbbell Suspensions with Brownian Motion. Vol. 8, pp. 1–90.
*Biswas, M.* and *Maity, C.:* Molecular Sieves as Polymerization Catalysts. Vol. 31, pp. 47–88.
*Block, H.:* The Nature and Application of Electrical Phenomena in Polymers. Vol. 33, pp. 93–167.
*Bodor, G.:* X-ray Line Shape Analysis. A. Means for the Characterization of Crystalline Polymers. Vol. 67, pp. 165–194.
*Böhm, L. L., Chmelíř, M., Löhr, G., Schmitt, B. J.* and *Schulz, G. V.:* Zustände und Reaktionen des Carbanions bei der anionischen Polymerisation des Styrols. Vol. 9, pp. 1–45.
*Bovey, F. A.* and *Tiers, G. V. D.:* The High Resolution Nuclear Magnetic Resonance Spectroscopy of Polymers. Vol. 3, pp. 139–195.
*Braun, J.-M.* and *Guillet, J. E.:* Study of Polymers by Inverse Gas Chromatography. Vol. 21, pp. 107–145.
*Breitenbach, J. W., Olaj, O. F.* und *Sommer, F.:* Polymerisationsanregung durch Elektrolyse. Vol. 9, pp. 47–227.

*Bresler, S. E.* and *Kazbekov, E. N.:* Macroradical Reactivity Studied by Electron Spin Resonance. Vol. 3, pp. 688–711.
*Bucknall, C. B.:* Fracture and Failure of Multiphase Polymers and Polymer Composites. Vol. 27, pp. 121–148.
*Burchard, W.:* Static and Dynamic Light Scattering from Branched Polymers and Biopolymers. Vol. 48, pp. 1–124.
*Bywater, S.:* Polymerization Initiated by Lithium and Its Compounds. Vol. 4, pp. 66–110.
*Bywater, S.:* Preparation and Properties of Star-branched Polymers. Vol. 30, pp. 89–116.

*Candau, S., Bastide, J.* and *Delsanti, M.:* Structural. Elastic and Dynamic Properties of Swollen Polymer Networks. Vol. 44, pp. 27–72.
*Carrick, W. L.:* The Mechanism of Olefin Polymerization by Ziegler-Natta Catalysts. Vol. 12, pp. 65–86.
*Casale, A.* and *Porter, R. S.:* Mechanical Synthesis of Block and Graft Copolymers. Vol. 17, pp. 1–71.
*Cerf, R.:* La dynamique des solutions de macromolecules dans un champ de vitesses. Vol. 1, pp. 382–450.
*Cesca, S., Priola, A.* and *Bruzzone, M.:* Synthesis and Modification of Polymers Containing a System of Conjugated Double Bonds. Vol. 32, pp. 1–67.
*Chiellini, E., Solaro R., Galli, G.* and *Ledwith, A.:* Pptically Active Synthetic Polymers Containing Pendant Carbazolyl Groups. Vol. 62, pp. 143–170.
*Cicchetti, O.:* Mechanisms of Oxidative Photodegradation and of UV Stabilization of Polyolefins. Vol. 7, pp. 70–112.
*Clark, D. T.:* ESCA Applied to Polymers. Vol. 24, pp. 125–188.
*Coleman, Jr., L. E.* and *Meinhardt, N. A.:* Polymerization Reactions of Vinyl Ketones. Vol. 1, pp. 159–179.
*Comper, W. D.* and *Preston, B. N.:* Rapid Polymer Transport in Concentrated Solutions. Vol. 55, pp. 105–152.
*Corner, T.:* Free Radical Polymerization – The Synthesis of Graft Copolymers. Vol. 62, pp. 95–142.
*Crescenzi, V.:* Some Recent Studies of Polyelectrolyte Solutions. Vol. 5, pp. 358–386.
*Crivello, J. V.:* Cationic Polymerization – Iodonium and Sulfonium Salt Photoinitiators, Vol. 62, pp. 1–48.

*Davydov, B. E.* and *Krentsel, B. A.:* Progress in the Chemistry of Polyconjugated Systems. Vol. 25, pp. 1–46.
*Dettenmaier, M.:* Intrinsic Crazes in Polycarbonate Phenomenology and Molecular Interpretation of a New Phenomenon. Vol. 52/53, pp. 57–104
*Dobb, M. G.* and *McIntyre, J. E.:* Properties and Applications of Liquid-Crystalline Main-Chain Polymers. Vol. 60/61, pp. 61–98.
*Döll, W.:* Optical Interference Measurements and Fracture Mechanics Analysis of Crack Tip Craze Zones. Vol. 52/53, pp. 105–168
*Dole, M.:* Calorimetric Studies of States and Transitions in Solid High Polymers. Vol. 2, pp. 221–274.
*Dorn, K., Hupfer, B.,* and *Ringsdorf, H.:* Polymeric Monolayers and Liposomes as Models for Biomembranes How to Bridge the Gap Between Polymer Science and Membrane Biology? Vol. 64, pp. 1–54.
*Dreyfuss, P.* and *Dreyfuss, M. P.:* Polytetrahydrofuran. Vol. 4, pp. 528–590.
*Drobnik, J.* and *Rypáček, F.:* Soluble Synthetic Polymers in Biological Systems. Vol. 57, pp. 1–50.
*Dröscher, M.:* Solid State Extrusion of Semicrystalline Copolymers. Vol. 47, pp. 120–138.
*Dušek, K.* and *Prins, W.:* Structure and Elasticity of Non-Crystalline Polymer Networks. Vol. 6, pp. 1–102.
*Duncan, R.* and *Kopeček, J.:* Soluble Synthetic Polymers as Potential Drug Carriers. Vol. 57, pp. 51–101.

*Eastham, A. M.:* Some Aspects of the Polymerization of Cyclic Ethers. Vol. 2, pp. 18–50.
*Ehrlich, P.* and *Mortimer, G. A.:* Fundamentals of the Free-Radical Polymerization of Ethylene. Vol. 7, pp. 386–448.

*Eisenberg, A.:* Ionic Forces in Polymers. Vol. 5, pp. 59–112.
*Elias, H.-G., Bareiss, R.* und *Watterson, J. G.:* Mittelwerte des Molekulargewichts und anderer Eigenschaften. Vol. 11, pp. 111–204.
*Elsner, G., Riekel, Ch.* and *Zachmann, H. G.:* Synchrotron Radiation Physics. Vol. 67, pp. 1–58.
*Elyashevich, G. K.:* Thermodynamics and Kinetics of Orientational Crystallization of Flexible-Chain Polymers. Vol. 43, pp. 207–246.
*Enkelmann, V.:* Structural Aspects of the Topochemical Polymerization of Diacetylenes. Vol. 63. pp. 91–136.

*Ferruti, P.* and *Barbucci, R.:* Linear Amino Polymers: Synthesis, Protonation and Complex Formation. Vol. 58, pp. 55–92
*Finkelmann, H.* and *Rehage, G.:* Liquid Crystal Side-Chain Polymers. Vol. 60/61, pp. 99–172.
*Fischer, H.:* Freie Radikale während der Polymerisation, nachgewiesen und identifiziert durch Elektronenspinresonanz. Vol. 5, pp. 463–530.
*Flory, P. J.:* Molecular Theory of Liquid Crystals. Vol. 59, pp. 1–36.
*Ford, W. T.* and *Tomoi, M.:* Polymer-Supported Phase Transfer Catalysts Reaction Mechanisms. Vol. 55, pp. 49–104.
*Fradet, A.* and *Maréchal, E.:* Kinetics and Mechanisms of Polyesterifications. I. Reactions of Diols with Diacids. Vol. 43, pp. 51–144.
*Friedrich, K.:* Crazes and Shear Bands in Semi-Crystalline Thermoplastics. Vol. 52/53, pp. 225–274
*Fujita, H.:* Diffusion in Polymer-Diluent Systems. Vol. 3, pp. 1–47.
*Funke, W.:* Über die Strukturaufklärung vernetzter Makromoleküle, insbesondere vernetzter Polyesterharze, mit chemischen Methoden. Vol. 4, pp. 157–235.

*Gal'braikh, L. S.* and *Rigovin, Z. A.:* Chemical Transformation of Cellulose. Vol. 14, pp. 87–130.
*Galli, G.* see Chiellini, E. Vol. 62, pp. 143–170.
*Gallot, B. R. M.:* Preparation and Study of Block Copolymers with Ordered Structures, Vol. 29, pp. 85–156.
*Gandini, A.:* The Behaviour of Furan Derivatives in Polymerization Reactions. Vol. 25, pp. 47–96.
*Gandini, A.* and *Cheradame, H.:* Cationic Polymerization. Initiation with Alkenyl Monomers. Vol. 34/35, pp. 1–289.
*Geckeler, K., Pillai, V. N. R.,* and *Mutter, M.:* Applications of Soluble Polymeric Supports. Vol. 39, pp. 65–94.
*Gerrens, H.:* Kinetik der Emulsionspolymerisation. Vol. 1, pp. 234–328.
*Ghiggino, K. P., Roberts, A. J.* and *Phillips, D.:* Time-Resolved Fluorescence Techniques in Polymer and Biopolymer Studies. Vol. 40, pp. 69–167.
*Goethals, E. J.:* The Formation of Cyclic Oligomers in the Cationic Polymerization of Heterocycles. Vol. 23, pp. 103–130.
*Graessley, W. W.:* The Etanglement Concept in Polymer Rheology. Vol. 16, pp. 1–179.
*Graessley, W. W.:* Entagled Linear, Branched and Network Polymer Systems. Molecular Theories. Vol. 47, pp. 67–117.
*Grebowicz, J.* see Wunderlich, B. Vol. 60/61, pp. 1–60.

*Hagihara, N., Sonogashira, K.* and *Takahashi, S.:* Linear Polymers Containing Transition Metals in the Main Chain. Vol. 41, pp. 149–179.
*Hasegawa, M.:* Four-Center Photopolymerization in the Crystalline State. Vol. 42, pp. 1–49.
*Hay, A. S.:* Aromatic Polyethers. Vol. 4, pp. 496–527.
*Hayakawa, R.* and *Wada, Y.:* Piezoelectricity and Related Properties of Polymer Films. Vol. 11, pp. 1–55.
*Heidemann, E.* and *Roth, W.:* Synthesis and Investigation of Collagen Model Peptides. Vol. 43, pp. 145–205.
*Heitz, W.:* Polymeric Reagents. Polymer Design, Scope, and Limitations. Vol. 23, pp. 1–23.
*Helfferich, F.:* Ionenaustausch. Vol. 1, pp. 329–381.
*Hendra, P. J.:* Laser-Raman Spectra of Polymers. Vol. 6, pp. 151–169.
*Hendrix, J.:* Position Sensitive "X-ray Detectors". Vol. 67, pp. 59–98.

*Henrici-Olivé, G.* und *Olivé, S.:* Kettenübertragung bei der radikalischen Polymerisation. Vol. 2, pp. 496–577.
*Henrici-Olivé, G.* und *Olivé, S.:* Koordinative Polymerisation an löslichen Übergangsmetall-Katalysatoren. Vol. 6, pp. 421–472.
*Henrici-Olivé, G.* and *Olivé, S.:* Oligomerization of Ethylene with Soluble Transition-Metal Catalysts. Vol. 15, pp. 1–30.
*Henrici-Olivé, G.* and *Olivé, S.:* Molecular Interactions and Macroscopic Properties of Polyacrylonitrile and Model Substances. Vol. 32, pp. 123–152.
*Henrici-Olivé, G.* and *Olivé, S.:* The Chemistry of Carbon Fiber Formation from Polyacrylonitrile. Vol. 51, pp. 1–60.
*Hermans, Jr., J., Lohr, D.* and *Ferro, D.:* Treatment of the Folding and Unfolding of Protein Molecules in Solution According to a Lattic Model. Vol. 9, pp. 229–283.
*Higashimura, T.* and *Sawamoto, M.:* Living Polymerization and Selective Dimerization: Two Extremes of the Polymer Synthesis by Cationic Polymerization. Vol. 62, pp. 49–94.
*Hoffman, A. S.:* Ionizing Radiation and Gas Plasma (or Glow) Discharge Treatments for Preparation of Novel Polymeric Biomaterials. Vol. 57, pp. 141–157.
*Holzmüller, W.:* Molecular Mobility, Deformation and Relaxation Processes in Polymers. Vol. 26, pp. 1–62.
*Hutchison, J.* and *Ledwith, A.:* Photoinitiation of Vinyl Polymerization by Aromatic Carbonyl Compounds. Vol. 14, pp. 49–86.

*Iizuka, E.:* Properties of Liquid Crystals of Polypeptides: with Stress on the Electromagnetic Orientation. Vol. 20, pp. 79–107.
*Ikada, Y.:* Characterization of Graft Copolymers. Vol. 29, pp. 47–84.
*Ikada, Y.:* Blood-Compatible Polymers. Vol. 57, pp. 103–140.
*Imanishi, Y.:* Synthese, Conformation, and Reactions of Cyclic Peptides. Vol. 20, pp. 1–77.
*Inagaki, H.:* Polymer Separation and Characterization by Thin-Layer Chromatography. Vol. 24, pp. 189–237.
*Inoue, S.:* Asymmetric Reactions of Synthetic Polypeptides. Vol. 21, pp. 77–106.
*Ise, N.:* Polymerizations under an Electric Field. Vol. 6, pp. 347–376.
*Ise, N.:* The Mean Activity Coefficient of Polyelectrolytes in Aqueous Solutions and Its Related Properties. Vol. 7, pp. 536–593.
*Isihara, A.:* Intramolecular Statistics of a Flexible Chain Molecule. Vol. 7, pp. 449–476.
*Isihara, A.:* Irreversible Processes in Solutions of Chain Polymers. Vol. 5, pp. 531–567.
*Isihara, A.* and *Guth, E.:* Theory of Dilute Macromolecular Solutions. Vol. 5, pp. 233–260.
*Iwatsuki, S.:* Polymerization of Quinodimethane Compounds. Vol. 58, pp. 93–120.

*Janeschitz-Kriegl, H.:* Flow Birefrigence of Elastico-Viscous Polymer Systems. Vol. 6, pp. 170–318.
*Jenkins, R.* and *Porter, R. S.:* Upertubed Dimensions of Stereoregular Polymers. Vol. 36, pp. 1–20.
*Jenngins, B. R.:* Electro-Optic Methods for Characterizing Macromolecules in Dilute Solution. Vol. 22, pp. 61–81.
*Johnston, D. S.:* Macrozwitterion Polymerization. Vol. 42, pp. 51–106.

*Kamachi, M.:* Influence of Solvent on Free Radical Polymerization of Vinyl Compounds. Vol. 38, pp. 55–87.
*Kaneko, M.* and *Yamada, A.:* Solar Energy Conversion by Functional Polymers. Vol. 55, pp. 1–48.
*Kawabata, S.* and *Kawai, H.:* Strain Energy Density Functions of Rubber Vulcanizates from Biaxial Extension. Vol. 24, pp. 89–124.
*Kennedy, J. P.* and *Chou, T.:* Poly(isobutylene-co-β-Pinene): A New Sulfur Vulcanizable, Ozone Resistant Elastomer by Cationic Isomerization Copolymerization. Vol. 21, pp. 1–39.
*Kennedy, J. P.* and *Delvaux, J. M.:* Synthesis, Characterization and Morphology of Poly(butadiene-g-Styrene). Vol. 38, pp. 141–163.
*Kennedy, J. P.* and *Gillham, J. K.:* Cationic Polymerization of Olefins with Alkylaluminium Initiators. Vol. 10, pp. 1–33.

*Kennedy, J. P.* and *Johnston, J. E.:* The Cationic Isomerization Polymerization of 3-Methyl-1-butene and 4-Methyl-1-pentene. Vol. 19, pp. 57–95.

*Kennedy, J. P.* and *Langer, Jr., A. W.:* Recent Advances in Cationic Polymerization. Vol. 3, pp. 508–580.

*Kennedy, J. P.* and *Otsu, T.:* Polymerization with Isomerization of Monomer Preceding Propagation. Vol. 7, pp. 369–385.

*Kennedy, J. P.* and *Rengachary, S.:* Correlation Between Cationic Model and Polymerization Reactions of Olefins. Vol. 14, pp. 1–48.

*Kennedy, J. P.* and *Trivedi, P. D.:* Cationic Olefin Polymerization Using Alkyl Halide — Alkylaluminium Initiator Systems. I. Reactivity Studies. II. Molecular Weight Studies. Vol. 28, pp. 83–151.

*Kennedy, J. P., Chang, V. S. C.* and *Guyot, A.:* Carbocationic Synthesis and Characterization of Polyolefins with Si–H and Si–Cl Head Groups. Vol. 43, pp. 1–50.

*Khoklov, A. R.* and *Grosberg, A. Yu.:* Statistical Theory of Polymeric Lyotropic Liquid Crystals. Vol. 41, pp. 53–97.

*Kissin, Yu. V.:* Structures of Copolymers of High Olefins. Vol. 15, pp. 91–155.

*Kitagawa, T.* and *Miyazawa, T.:* Neutron Scattering and Normal Vibrations of Polymers. Vol. 9, pp. 335–414.

*Kitamaru, R.* and *Horii, F.:* NMR Approach to the Phase Structure of Linear Polyethylene. Vol. 26, pp. 139–180.

*Knappe, W.:* Wärmeleitung in Polymeren. Vol. 7, pp. 477–535.

*Koenig, J. L.:* Fourier Transforms Infrared Spectroscopy of Polymers, Vol. 54, pp. 87–154.

*Kolařík, J.:* Secondary Relaxations in Glassy Polymers: Hydrophilic Polymethacrylates and Polyacrylates: Vol. 46, pp. 119–161.

*Koningsveld, R.:* Preparative and Analytical Aspects of Polymer Fractionation. Vol. 7.

*Kovacs, A. J.:* Transition vitreuse dans les polymers amorphes. Etude phénoménologique. Vol. 3, pp. 394–507.

*Krässig, H. A.:* Graft Co-Polymerization of Cellulose and Its Derivatives. Vol. 4, pp. 111–156.

*Kramer, E. J.:* Microscopic and Molecular Fundamentals of Crazing. Vol. 52/53, pp. 1–56

*Kraus, G.:* Reinforcement of Elastomers by Carbon Black. Vol. 8, pp. 155–237.

*Kreutz, W.* and *Welte, W.:* A General Theory for the Evaluation of X-Ray Diagrams of Biomembranes and Other Lamellar Systems. Vol. 30, pp. 161–225.

*Krimm, S.:* Infrared Spectra of High Polymers. Vol. 2, pp. 51–72.

*Kuhn, W., Ramel, A., Walters, D. H., Ebner, G.* and *Kuhn, H. J.:* The Production of Mechanical Energy from Different Forms of Chemical Energy with Homogeneous and Cross-Striated High Polymer Systems. Vol. 1, pp. 540–592.

*Kunitake, T.* and *Okahata, Y.:* Catalytic Hydrolysis by Synthetic Polymers. Vol. 20, pp. 159–221.

*Kurata, M.* and *Stockmayer, W. H.:* Intrinsic Viscosities and Unperturbed Dimensions of Long Chain Molecules. Vol. 3, pp. 196–312.


*Ledwith, A.* and *Sherrington, D. C.:* Stable Organic Cation Salts: Ion Pair Equilibria and Use in Cationic Polymerization. Vol. 19, pp. 1–56.

*Ledwith, A.* see Chiellini, E. Vol. 62, pp. 143–170.

*Lee, C.-D. S.* and *Daly, W. H.:* Mercaptan-Containing Polymers. Vol. 15, pp. 61–90.

*Lindberg, J. J.* and *Hortling, B.:* Cross Polarization — Magic Angle Spinning NMR Studies of Carbohydrates and Aromatic Polymers. Vol. 66, pp. 1–22.

*Lipatov, Y. S.:* Relaxation and Viscoelastic Properties of Heterogeneous Polymeric Compositions. Vol. 22, pp. 1–59.

*Lipatov, Y. S.:* The Iso-Free-Volume State and Glass Transitions in Amorphous Polymers: New Development of the Theory. Vol. 26, pp. 63–104.

*Lustoň, J.* and *Vašš, F.:* Anionic Copolymerization of Cyclic Ethers with Cyclic Anhydrides. Vol. 56, pp. 91–133.


*Mano, E. B.* and *Coutinho, F. M. B.:* Grafting on Polyamides. Vol. 19, pp. 97–116.

*Mark, J. E.:* The Use of Model Polymer Networks to Elucidate Molecular Aspects of Rubberlike Elasticity. Vol. 44, pp. 1–26.

*Maser, F.*, *Bode, K.*, *Pillai, V. N. R.* and *Mutter, M.*: Conformational Studies on Model Peptides. Their Contribution to Synthetic, Structural and Functional Innovations on Proteins. Vol. 65, pp. 177–214.
*McIntyre, J. E.* see Dobb, M. G. Vol. 60/61, pp. 61–98.
*Meerwall v., E., D.*: Self-Diffusion in Polymer Systems, Measured with Field-Gradient Spin Echo NMR Methods, Vol. 54, pp. 1–29.
*Mengoli, G.*: Feasibility of Polymer Film Coating Through Electroinitiated Polymerization in Aqueous Medium. Vol. 33, pp. 1–31.
*Meyerhoff, G.*: Die viscosimetrische Molekulargewichtsbestimmung von Polymeren. Vol. 3, pp. 59–105.
*Millich, F.*: Rigid Rods and the Characterization of Polyisocyanides. Vol. 19, pp. 117–141.
*Möller, M.*: Cross Polarization — Magic Angle Sample Spinning NMR Studies. With Respect to the Rotational Isomeric States of Saturated Chain Molecules. Vol. 66, pp. 59–80.
*Morawetz, H.*: Specific Ion Binding by Polyelectrolytes. Vol. 1, pp. 1–34.
*Morin, B. P.*, *Breusova, I. P.* and *Rogovin, Z. A.*: Structural and Chemical Modifications of Cellulose by Graft Copolymerization. Vol. 42, pp. 139–166.
*Mulvaney, J. E.*, *Oversberger, C. C.* and *Schiller, A. M.*: Anionic Polymerization. Vol. 3, pp. 106–138.


*Nakase, Y.*, *Kurijama, I.* and *Odajima, A.*: Analysis of the Fine Structure of Poly(Oxymethylene) Prepared by Radiation-Induced Polymerization in the Solid State. Vol. 65, pp. 79–134.
*Neuse, E.*: Aromatic Polybenzimidazoles. Syntheses, Properties, and Applications. Vol. 47, pp. 1–42.


*Ober, Ch. K.*, *Jin, J.-I.* and *Lenz, R. W.*: Liquid Crystal Polymers with Flexible Spacers in the Main Chain. Vol. 59, pp. 103–146.
*Okubo, T.* and *Ise, N.*: Synthetic Polyelectrolytes as Models of Nucleic Acids and Esterases. Vol. 25, pp. 135–181.
*Osaki, K.*: Viscoelastic Properties of Dilute Polymer Solutions. Vol. 12, pp. 1–64.
*Oster, G.* and *Nishijima, Y.*: Fluorescence Methods in Polymer Science. Vol. 3, pp. 313–331.
*Overberger, C. G.* and *Moore, J. A.*: Ladder Polymers. Vol. 7, pp. 113–150.


*Papkov, S. P.*: Liquid Crystalline Order in Solutions of Rigid-Chain Polymers. Vol. 59, pp. 75–102.
*Patat, F.*, *Killmann, E.* und *Schiebener, C.*: Die Absorption von Makromolekülen aus Lösung. Vol. 3, pp. 332–393.
*Patterson, G. D.*: Photon Correlation Spectroscopy of Bulk Polymers. Vol. 48, pp. 125–159.
*Penczek, S.*, *Kubisa, P.* and *Matyjaszewski, K.*: Cationic Ring-Opening Polymerization of Heterocyclic Monomers. Vol. 37, pp. 1–149.
*Peticolas, W. L.*: Inelastic Laser Light Scattering from Biological and Synthetic Polymers. Vol. 9, pp. 285–333.
*Petropoulos, J. H.*: Membranes with Non-Homogeneous Sorption Properties. Vol. 64, pp. 85–134.
*Pino, P.*: Optically Active Addition Polymers. Vol. 4, pp. 393–456.
*Pitha, J.*: Physiological Activities of Synthetic Analogs of Polynucleotides. Vol. 50, pp. 1–16.
*Platé, N. A.* and *Noak, O. V.*: A Theoretical Consideration of the Kinetics and Statistics of Reactions of Functional Groups of Macromolecules. Vol. 31, pp. 133–173.
*Platé, N. A.* see Shibaev, V. P. Vol. 60/61, pp. 173–252.
*Plesch, P. H.*: The Propagation Rate-Constants in Cationic Polymerisations. Vol. 8, pp. 137–154.
*Porod, G.*: Anwendung und Ergebnisse der Röntgenkleinwinkelstreuung in festen Hochpolymeren. Vol. 2, pp. 363–400.
*Pospíšil, J.*: Transformations of Phenolic Antioxidants and the Role of Their Products in the Long-Term Properties of Polyolefins. Vol. 36, pp. 69–133.
*Postelnek, W.*, *Coleman, L. E.*, and *Lovelace, A. M.*: Fluorine-Containing Polymers. I. Fluorinated Vinyl Polymers with Functional Groups, Condensation Polymers, and Styrene Polymers. Vol. 1, pp. 75–113.

*Queslel, J. P.* and *Mark, J. E.:* Molecular Interpretation of the Moduli of Elastomeric Polymer Networks of Know Structure. Vol. 65, pp. 135–176.

*Rehage, G.* see Finkelmann, H. Vol. 60/61, pp. 99–172.
*Rempp, P. F.* and *Franta, E.:* Macromonomers: Synthesis, Characterization and Applications. Vol. 58, pp. 1–54.
*Rempp, P., Herz, J.*, and *Borchard, W.:* Model Networks. Vol. 26, pp. 107–137.
*Rigbi, Z.:* Reinforcement of Rubber by Carbon Black. Vol. 36, pp. 21–68.
*Rogovin, Z. A.* and *Gabrielyan, G. A.:* Chemical Modifications of Fibre Forming Polymers and Copolymers of Acrylonitrile. Vol. 25, pp. 97–134.
*Roha, M.:* Ionic Factors in Steric Control. Vol. 4, pp. 353–392.
*Roha, M.:* The Chemistry of Coordinate Polymerization of Dienes. Vol. 1, pp. 512–539.

*Safford, G. J.* and *Naumann, A. W.:* Low Frequency Motions in Polymers as Measured by Neutron Inelastic Scattering. Vol. 5, pp. 1–27.
*Sauer, J. A.* and *Chen, C. C.:* Crazing and Fatigue Behavior in One and Two Phase Glassy Polymers. Vol. 52/53, pp. 169–224
*Sawamoto, M.* see Higashimura, T. Vol. 62, pp. 49–94.
*Schuerch, C.:* The Chemical Synthesis and Properties of Polysaccharides of Biomedical Interest. Vol. 10, pp. 173–194.
*Schulz, R. C.* und *Kaiser, E.:* Synthese und Eigenschaften von optisch aktiven Polymeren. Vol. 4, pp. 236–315.
*Seanor, D. A.:* Charge Transfer in Polymers. Vol. 4, pp. 317–352.
*Semerak, S. N.* and *Frank, C. W.:* Photophysics of Excimer Formation in Aryl Vinyl Polymers, Vol. 54, pp. 31–85.
*Seidl, J., Malinský, J., Dušek, K.* und *Heitz, W.:* Makroporöse Styrol-Divinylbenzol-Copolymere und ihre Verwendung in der Chromatographie und zur Darstellung von Ionenaustauschern. Vol. 5, pp. 113–213.
*Semjonow, V.:* Schmelzviskositäten hochpolymerer Stoffe. Vol. 5, pp. 387–450.
*Semlyen, J. A.:* Ring-Chain Equilibria and the Conformations of Polymer Chains. Vol. 21, pp. 41–75.
*Sharkey, W. H.:* Polymerizations Through the Carbon-Sulphur Double Bond. Vol. 17, pp. 73–103.
*Shibaev, V. P.* and *Platé, N. A.:* Thermotropic Liquid-Crystalline Polymers with Mesogenic Side Groups. Vol. 60/61, pp. 173–252.
*Shimidzu, T.:* Cooperative Actions in the Nucleophile-Containing Polymers. Vol. 23, pp. 55–102.
*Shutov, F. A.:* Foamed Polymers Based on Reactive Oligomers, Vol. 39, pp. 1–64.
*Shutov, F. A.:* Foamed Polymers. Cellular Structure and Properties. Vol. 51, pp. 155–218.
*Siesler, H. W.:* Rheo-Optical Fourier-Transform Infrared Spectroscopy: Vibrational Spectra and Mechanical Properties of Polymers. Vol. 65, pp. 1–78.
*Silvestri, G., Gambino, S.*, and *Filardo, G.:* Electrochemical Production of Initiators for Polymerization Processes. Vol. 38, pp. 27–54.
*Sixl, H.:* Spectroscopy of the Intermediate States of the Solid State Polymerization Reaction in Diacetylene Crystals. Vol. 63, pp. 49–90.
*Slichter, W. P.:* The Study of High Polymers by Nuclear Magnetic Resonance. Vol. 1, pp. 35–74.
*Small, P. A.:* Long-Chain Branching in Polymers. Vol. 18.
*Smets, G.:* Block and Graft Copolymers. Vol. 2, pp. 173–220.
*Smets, G.:* Photochromic Phenomena in the Solid Phase. Vol. 50, pp. 17–44.
*Sohma, J.* and *Sakaguchi, M.:* ESR Studies on Polymer Radicals Produced by Mechanical Destruction and Their Reactivity. Vol. 20, pp. 109–158.
*Solaro, R.* see Chiellini, E. Vol. 62, pp. 143–170.
*Sotobayashi, H.* und *Springer, J.:* Oligomere in verdünnten Lösungen. Vol. 6, pp. 473–548.
*Sperati, C. A.* and *Starkweather, Jr., H. W.:* Fluorine-Containing Polymers. II. Polytetrafluoroethylene. Vol. 2, pp. 465–495.
*Spiess, H. W.:* Deuteron NMR — A new Tool for Studying Chain Mobility and Orientation in Polymers. Vol. 66, pp. 23–58.
*Sprung, M. M.:* Recent Progress in Silicone Chemistry. I. Hydrolysis of Reactive Silane Intermediates, Vol. 2, pp. 442–464.

*Stahl, E.* and *Brüderle, V.:* Polymer Analysis by Thermofractography. Vol. 30, pp. 1–88.
*Stannett, V. T., Koros, W. J., Paul, D. R., Lonsdale, H. K.,* and *Baker, R. W.:* Recent Advances in Membrane Science and Technology. Vol. 32, pp. 69–121.
*Staverman, A. J.:* Properties of Phantom Networks and Real Networks. Vol. 44, pp. 73–102.
*Stauffer, D., Coniglio, A.* and *Adam, M.:* Gelation and Critical Phenomena. Vol. 44, pp. 103–158.
*Stille, J. K.:* Diels-Alder Polymerization. Vol. 3, pp. 48–58.
*Stolka, M.* and *Pai, D.:* Polymers with Photoconductive Properties. Vol. 29, pp. 1–45.
*Stuhrmann, H.:* Resonance Scattering in Macromolecular Structure Research. Vol. 67, pp. 123–164.
*Subramanian, R. V.:* Electroinitiated Polymerization on Electrodes. Vol. 33, pp. 35–58.
*Sumitomo, H.* and *Hashimoto, K.:* Polyamides as Barrier Materials. Vol. 64, pp. 55–84.
*Sumitomo, H.* and *Okada, M.:* Ring-Opening Polymerization of Bicyclic Acetals, Oxalactone, and Oxalactam. Vol. 28, pp. 47–82.
*Szegö, L.:* Modified Polyethylene Terephthalate Fibers. Vol. 31, pp. 89–131.
*Szwarc, M.:* Termination of Anionic Polymerization. Vol. 2, pp. 275–306.
*Szwarc, M.:* The Kinetics and Mechanism of N-carboxy-α-amino-acid Anhydride (NCA) Polymerization to Poly-amino Acids. Vol. 4, pp. 1–65.
*Szwarc, M.:* Thermodynamics of Polymerization with Special Emphasis on Living Polymers. Vol. 4, pp. 457–495.
*Szwarc, M.:* Living Polymers and Mechanisms of Anionic Polymerization. Vol. 49, pp. 1–175.


*Takahashi, A.* and *Kawaguchi, M.:* The Structure of Macromolecules Adsorbed on Interfaces. Vol. 46, pp. 1–65.
*Takemoto, K.* and *Inaki, Y.:* Synthetic Nucleic Acid Analogs. Preparation and Interactions. Vol. 41, pp. 1–51.
*Tani, H.:* Stereospecific Polymerization of Aldehydes and Epoxides. Vol. 11, pp. 57–110.
*Tate, B. E.:* Polymerization of Itaconic Acid and Derivatives. Vol. 5, pp. 214–232.
*Tazuke, S.:* Photosensitized Charge Transfer Polymerization. Vol. 6, pp. 321–346.
*Teramoto, A.* and *Fujita, H.:* Conformation-dependent Properties of Synthetic Polypeptides in the Helix-Coil Transition Region. Vol. 18, pp. 65–149.
*Theocaris, P. S.:* The Mesophase and its Influence on the Mechanical Behavior of Composites. Vol. 66, pp. 149–188.
*Thomas, W. M.:* Mechanismus of Acrylonitrile Polymerization. Vol. 2, pp. 401–441.
*Tobolsky, A. V.* and *DuPré, D. B.:* Macromolecular Relaxation in the Damped Torsional Oscillator and Statistical Segment Models. Vol. 6, pp. 103–127.
*Tosi, C.* and *Ciampelli, F.:* Applications of Infrared Spectroscopy to Ethylene-Propylene Copolymers. Vol. 12, pp. 87–130.
*Tosi, C.:* Sequence Distribution in Copolymers: Numerical Tables. Vol. 5, pp. 451–462.
*Tsuchida, E.* and *Nishide, H.:* Polymer-Metal Complexes and Their Catalytic Activity. Vol. 24, pp. 1–87.
*Tsuji, K.:* ESR Study of Photodegradation of Polymers. Vol. 12, pp. 131–190.
*Tsvetkov, V.* and *Andreeva, L.:* Flow and Electric Birefringence in Rigid-Chain Polymer Solutions. Vol. 39, pp. 95–207.
*Tuzar, Z., Kratochvíl, P.,* and *Bohdanecký, M.:* Dilute Solution Properties of Aliphatic Polyamides. Vol. 30, pp. 117–159.


*Uematsu, I.* and *Uematsu, Y.:* Polypeptide Liquid Crystals. Vol. 59, pp. 37–74.


*Valvassori, A.* and *Sartori, G.:* Present Status of the Multicomponent Copolymerization Theory. Vol. 5, pp. 28–58.
*Viovy, J. L.* and *Monnerie, L.:* Fluorescence Anisotropy Technique Using Synchrotron Radiation as a Powerful Means for Studying the Orientation Correlation Functions of Polymer Chains. Vol. 67, pp. 99–122.
*Voigt-Martin, I.:* Use of Transmission Electron Microscopy to Obtain Quantitative Information About Polymers. Vol. 67, pp. 195–218.
*Voorn, M. J.:* Phase Separation in Polymer Solutions. Vol. 1, pp. 192–233.

*Ward, I. M.:* Determination of Molecular Orientation by Spectroscopic Techniques. Vol. 66, pp. 81–116.
*Werber, F. X.:* Polymerization of Olefins on Supported Catalysts. Vol. 1, pp. 180–191.
*Wichterle, O., Šebenda, J.,* and *Králiček, J.:* The Anionic Polymerization of Caprolactam. Vol. 2, pp. 578–595.
*Wilkes, G. L.:* The Measurement of Molecular Orientation in Polymeric Solids. Vol. 8, pp. 91–136.
*Williams, G.:* Molecular Aspects of Multiple Dielectric Relaxation Processes in Solid Polymers. Vol. 33, pp. 59–92.
*Williams, J. G.:* Applications of Linear Fracture Mechanics. Vol. 27, pp. 67–120.
*Wöhrle, D.:* Polymere aus Nitrilen. Vol. 10, pp. 35–107.
*Wöhrle, D.:* Polymer Square Planar Metal Chelates for Science and Industry. Synthesis, Properties and Applications. Vol. 50, pp. 45–134.
*Wolf, B. A.:* Zur Thermodynamik der enthalpisch und der entropisch bedingten Entmischung von Polymerlösungen. Vol. 10, pp. 109–171.
*Woodward, A. E.* and *Sauer, J. A.:* The Dynamic Mechanical Properties of High Polymers at Low Temperatures. Vol. 1, pp. 114–158.
*Wunderlich, B.:* Crystallization During Polymerization. Vol. 5, pp. 568–619.
*Wunderlich, B.* and *Baur, H.:* Heat Capacities of Linear High Polymers. Vol. 7, pp. 151–368.
*Wunderlich, B.* and *Grebowicz, J.:* Thermotropic Mesophases and Mesophase Transitions of Linear, Flexible Macromolecules. Vol. 60/61, pp. 1–60.
*Wrasidlo, W.:* Thermal Analysis of Polymers. Vol. 13, pp. 1–99.

*Yamashita, Y.:* Random and Black Copolymers by Ring-Opening Polymerization. Vol. 28, pp. 1–46.
*Yamazaki, N.:* Electrolytically Initiated Polymerization. Vol. 6, pp. 377–400.
*Yamazaki, N.* and *Higashi, F.:* New Condensation Polymerizations by Means of Phosphorus Compounds. Vol. 38, pp. 1–25.
*Yokoyama, Y.* and *Hall, H. K.:* Ring-Opening Polymerization of Atom-Bridged and Bond-Bridged Bicyclic Ethers, Acetals and Orthoesters. Vol. 42, pp. 107–138.
*Yoshida, H.* and *Hayashi, K.:* Initiation Process of Radiation-induced Ionic Polymerization as Studied by Electron Spin Resonance. Vol. 6, pp. 401–420.
*Young, R. N., Quirk, R. P.* and *Fetters, L. J.:* Anionic Polymerizations of Non-Polar Monomers Involving Lithium. Vol. 56, pp. 1–90.
*Yuki, H.* and *Hatada, K.:* Stereospecific Polymerization of Alpha-Substituted Acrylic Acid Esters. Vol. 31, pp. 1–45.

*Zachmann, H. G.:* Das Kristallisations- und Schmelzverhalten hochpolymerer Stoffe. Vol. 3, pp. 581–687.
*Zakharov, V. A., Bukatov, G. D.,* and *Yermakov, Y. I.:* On the Mechanism of Olifin Polymerization by Ziegler-Natta Catalysts. Vol. 51, pp. 61–100.
*Zambelli, A.* and *Tosi, C.:* Stereochemistry of Propylene Polymerization. Vol. 15, pp. 31–60.
*Zucchini, U.* and *Cecchin, G.:* Control of Molecular-Weight Distribution in Polyolefins Synthesized with Ziegler-Natta Catalytic Systems. Vol. 51, pp. 101–154.

# Subject Index

Absolute intensity 42, 148
Absorption by a sample 144
— coefficient 127
— edge 127, 131
Accelerator, linear 3
Amorphous material 138
Ancillary equipment 34
Angular dispersive cameras 24
— — method 22
— distribution 23
Anisotropy as function of temperature 114
— time dependence 104
Annealing above the crystallization temperature 41
Anode wire 31
Anomalous dispersion 24, 128, 133
Anthracene 109
Aperture slits 26, 29
Apoferritin 150
Argand diagram 130
Arrhenius plot 119
Asymmetry factor 27
Athermal process 46
Azimutal half-width 39

Backgammon detector 73
Bacterial thermoplastic 41
Band width 21
Barium 128
Basic scattering functions 135, 154
Beam 3
— compression 27
— expansion 27
Beamstop 26
Bendler and Yaris model 103
Bessel function 103
Binary mixture 139
Biological macromolecules 3
3-Bond motions 102
Bonse Hart camera 30
Bremsstrahlspektrum 20
Brightness of radiation source 17
— — synchrotron radiation source 21
Brownian motion 114

Bulk polymers 100, 104, 114
Bunch of electrons 3, 7
—, revolution time of one 9

Caesium 151
Calcium 152
CAMAC standard 92
Cathode 31
Cauchy function 178, 179
Chain orientation 39
Charge coupled devices (CCD) 89
Charge-division read-out 71
Chemical reactions 37, 54
— stain 196
Coherent elastic scattering 212
Compositional fluctuations 46
Compression 27
Conducting polymers 146, 160
Connectivity of labelled polystyrene 112
Continuous spectrum 109
Contracting muscle 91
Contrast 147
— variation 137, 151
Copolyester 51
Correlation functions 100, 212
Counting rates 108
Count rate limitation 33
Crazing 46
Crystalline particle, diameter size distribution 188
— — size 169, 175
— — — distribution 186
— — number size distribution 187
Crystallinity 203
—, degree of 40, 49
Crystallite size 181
Crystallization 47, 209
—, isothermal 49
—, kinetics of 41, 51
Crystallization time 38, 39
Curve-fitting 110, 112
Cytochrome oxidase 155

Data acquisition systems 31, 91

Deconvolution 166, 186
Degree of crystallinity 40, 49
Delay-line principle 31
— read-out 69
Demagnification of source point 25
Demagnified source size 30
Density difference 43
Desoxyribonucleic acid (DNA) 159
Diameter size distribution of crystalline
    particle 188
Diffuse scattering 29
— term 112
Dilatometry 47
Dilute solutions 110
Dipalmitoyl phosphatidyl chol in
    (DPPC) 155
Dipole magnets 11
Distortion broadening 167
— coefficient 168
Distribution of relaxation times 104
Double focussing, mirror-monochromator
    cameras 25
— monochromator camera 24

EDX (energy dispersive analysis of
    X-rays) 214, 215, 216
EELS (electron energy loss
    spectroscopy) 214, 216
Elastic scattering 125
Electric vector of emitted light, polarized 12
Electron beam damage 214, 216
— current 15
— diffraction 212
— microscopy 196, 203, 204
— —, analytical 214
Elemental analysis 215
EMBL 140
Emitted power 11
Energy-dispersive camera 23
— — method 22
—, total 17
Erbium 155, 156
ESR 117
Europium 128
EXAFS 130
Experimental windows 119
Extended amorphous chains 183

FAD 120
— apparatus 106
Fankuchen cut 27
Ferritin 149, 152
Figure of merit (FOM) 23
Filling process 7
Fine grid detector 80, 82
Flash lamp 108
Flexibility of labelled polystyrene 112
Float glass 27

Fluorescence 145
— anisotropy decay 105
— depolymerization 21
— yield 146
Fluorescent labelling 109
Focal length 26
— spot size 26
Focussing of source point 25
Form factor 179
Fourier analysis 181, 182
— coefficients 175, 185
— infrared spectroscopy 48
— method 174, 175
— transformation 212
— transforms 166
Friction coefficient 116
Friedel's law 132
Fundamental anisotropy $r_o$ 105

Gas amplification 95
— — A 62
Gas-filled detector, one-dimensional 31
— — systems 60
Gaussian function 178, 179
— ratio 186
Ge (111) 27
Glass transition 118
Grazing angle 27
Guard slit 26, 29
Gyration, dispersion of radius of 149
—, radius of 150, 159

Hall and Helfand model 103
Hankel transformations 136
HASYLAB 144
Heavy atom derivative 133
Helical wigglers 11
Hemoglobin 148, 153, 154
Higher harmonics 27
High pressure 23
Hook effect 181

Inhomogeneous strain 184
Integral breadth 167, 179
— — methods 173
Intensity distribution 19
Interference wiggler 11
Intrachain connectivity 116
Iodine 161
Ionization energy 150
Ionomers 158
Iron 148, 151, 153
Isomerization, cis/trans of polyacetylene 53
Isomorphous replacement 132
Isothermal crystallization 38, 196, 199, 203,
    204, 207, 212
— thickening 200, 204
Iterative reconvolution 107

Jones and Stockmayer model  103

Kerr effect  105
Kramers-Kronig relation  128
Kronig structure  130

Labelled lhains  110—114
— polystyrene  109
Lambda method  134
Lamella, bended  39
Lattice distortion  167, 175, 181, 183, 184
Linear accelerator  3
— position sensitive detector  26
— photodiode detector array  46
Line-broadening analysis  173
Line profile  178
Lipid bilayer  155
Liquid crystalline state  51
Local dynamics  101
— reorientation processes  120
Long period  38, 39, 43
Loss term  112

Macromolecules in solution  135
Macroscopic viscoelasticity, elementary process of  120
Main-chain orientation relaxation  114
Master equation  102
Melt crystallization  41
Melting  54
—-recrystallization process  46
— under isothermal conditions  54
Membranes  155
Microstrain  168, 175—177, 183
— function  178
— distortion  174
—, relative mean-square  168
Micro-diffraction  214
Mirror  27
—, segmented  27
Miscibility gap  46
Modulation-transfer function (MTF)  86
Molecular orbitals  131
— tilt angle  209
Monochromator  27
— beam  23
— crystals  141
—, double  140
—, resolution of  142
—, triangular  26
Monoclinic modification  53
Morphology of crystallite  196, 203
Multipole  136
— wiggler  11
Multi-wavelength methods  133
Multiwire proportional chamber  75, 76

Neutron scattering  151, 156, 212

NMR  117, 120
Noncrystalline regions  43
Non-linear least square  107
Nuclear resonance  130
Nucleic acids  158
Number size distribution of crystalline particles  187
Nylon12  46

Oligo-butyleneterephthalate  46
Oligo-oxytetramethylene  46
One-dimension diffusion  102
Optical elements  26
— theorem  124, 146
Optimal thickness  145
Orthorhombic crystal  209
Orientation auto correlation functions (OACF)  101
Oxyhemoglobin  155

Paracrystalline disorder  184
— distortion  169, 174, 175, 176
Paracrystallinity  183
Partial distribution function  138
— melting  43
Particle acceleration  3
— current, time structure  7
— size  169, 177, 184
— — broadening  167
— — coefficient  185
— — distribution  185
Parvalbumin  152
Period wiggler, single  11
PET, strecking behavior  46
α-Phase  49
β-—  49
Phase problem  133
— separation  46
Phosphorus  144
Photo-Diode-Array (PDA)  90
Photoelectric absorption  126
Photoelectrons  131
Photon energy  17
— — distribution  22
Pinhole camera  26
Platinum-coated mirrors  28
Polarization  12, 140
—, circular  15
—, linear  15
Polarized light, elliptically  15
Polyacetylene  53, 54, 160
—, decomposition of  53
Polyamides  50
Polybutadiene  46, 114
Polychromatic beam  22
Polyelectrolytes  158
Polyethylacrylate  46
Polyethylene  41, 42, 53

—, branched 204
—-naphthalene-2,6-dicarboxylate 51
—, oriented 43
—, orthorhombic 53, 209
— terephthalate 38, 43, 54
— —, highly oriented 41
Poly-β-hydroxybutyrate (PHB) 41
Polyisobutylene 48
Polymer blends 46
— dynamics 21, 100
—, radiation damage of 196
Polymethylmethacrylate 46
Polypropylene 49
—, isotactic 49
—, oriented 41
Polystyrene 110
—, labelled 109
Polystyrole 46
Polytetrafluorethylene 175
Polyvinylidenefluoride 46
Position-sensitive area detectors 149
— detectors 60
— — —, one-dimensional 31
— — —, two-dimensional 33
Power, emitted 15
Primary lamellae 41
Proportional chamber 62
— counter detector 34
— —, area detector 24
Proteins 147
— structure 133
Pulse fluorometry 105
— shape 66, 108

Quadrupole magnet 9
Quartz mirror 26, 27
Quenching 196, 197, 204

Radial distribution function 214
Radiation 15
— damage of polymer 196
—, polarized 15
—, total 15
Radius of gyration of cross section of DNA 159
Raman spectroscopy 196
Rare earth ions 129
RC-read-out detector 74
Read-out technique for two-dimensional systems 77
Recrystallization 42, 43
Redox centres 155
Reduced $\chi^2$ 109
Reduced displacement 168
Reflection, total 27
Refractive index 143
Relative mean-square fluctuation 175
Resolution 28

Resonance neutron scattering 128, 130
— scattering 124
— — instruments 140
Revolution frequency 12
— time of one bunch 9
Ribosome 138, 148, 159
Rotational diffusion models 102
Rouse model 102
Rubber 48

Sampling 181
Scanning electron microscopy 216
— transmission electron microscopy (STEM) 214, 215
Scattering amplitude 132
— function 212
— intensity 132
— length 125
— power 43
— — Q 40
Segmented block copolymers 46
Selenium 148
Sextupole magnet 9
Si (220) 27
Silicon 144
— detector 23
— diode-array target 85
—-intensifier-tanget-Tube (SIT-tube) 84
Single-line profile analysis 183
— photon counting 106
Small-angle scattering 38, 41
— scattering camera 23
— — experiments 26
— — resolution 29
Smectic modification 49
Solid-state phase transition 47
Solution crystallization 41
Source point, focussing of 25
— size 31
Space charge effect 64
Spatial resolution 69
Spectral distribution 12, 13
Spherical harmonics 101, 136
Spinodal decomposition 41, 46
STEM (Scanning transmission electron microscopy) 214, 215
Stokes deconvolution 187
Storage ring 7, 9
Strain 169
—, average 184, 185
— distribution 184, 186
—, inhomogeneous 184
— variance 184
Stress-induced crazes 46
Stretch-induced tansformation 53
Stretching device 37
— experiments 37
Structural broadening 166, 184

Subject Index

Styrole/butadiene blockcopolymers 46
Sulfur 144
Surface replica 212
— roughness 28
Svergun method 137
Synchrotron 7, 9
Synchrotron radiation 3, 13, 17, 19, 20, 21, 60, 107, 140, 149

Temperature effects 117
Terbium 152, 153
Tetramethylammonium 159
Thermoplast 158
Time frame generator 32, 93
Time-temperature superposition 118
Time-to-amplitude converter 106
Time-to-digital converter 32
Time-resolved measurements 91
Topological interchain effects 116
Total cross section 126
— reflection 143
Triangular monochromator 26
Truncation 181, 186
TV-tube detector 60
Two-dimensional detectors 33
— systems 77

Undulators 10, 11, 19

Vacuum ultraviolet spectroscopy 3
Valeur et al. model 102
Vertical divergence 30
Vidicon tube 33
— — based detectors 83
Voigt curve 185
— function 178—180
— —, full width at half height 180

Warren-Averbach method 184, 185
— plot 168, 175, 185, 187, 189
Wavelength cutoff 28
— —, critical 13, 19
— —, distribution 13
— — range 19
Wide-angle scattering resolution 30
Wigglers 10, 11, 19
Williams-Watts distribution 104
Wire-per-wire read-out 79
WLF curve 118

XANES 130
— region 139
X-ray diffraction 22
— lines, characteristic 20
— scattering 3, 196, 209, 212
— tubes 20, 21
— tube radiation 21
— range, for polymer research 19

# Crazing in Polymers

Editor: **H. H. Kausch**
With contributions by numerous experts

1983. 234 figures, 12 tables. IX, 347 pages
(Advances in Polymer Science, Fortschritte der Hochpolymeren-Forschung, Volume 52/53)
ISBN 3-540-12571-X

**Contents/Information:**

*E. J. Kramer:* **Microscopic and Molecular Fundamentals of Crazing.** The first chapter considers crazes produced in air. The electron beam imaging technique developed in Ithaca is shown to be a powerful tool to determine the micromechanics and microstructure of crazes. (133 references)

*M. Dettenmaier:* **Intrinsic Crazes in Polycarbonate: Phenomenology and Molecular Interpretation of a New Phenomenon.** The second chapter demonstrates that the intrinsic crazing of polycarbonate is related to the existence of a general mode of cavitational plasticity in glassy polymers. (193 references)

*W. Döll:* **Optical Interference Measurements and Fracture Mechanics Analysis of Crazes.** The third chapter describes the optical interference measurement providing considerable insight into the role of crazes in deformation and fracture of amorphous polymers. (142 references)

*J. A. Sauer, C. C. Chen:* **Crazing and Fatigue in One- and Two-Phase Glassy Polymers.** The fourth chapter investigates deformation and fracture mechanisms in glassy polymer systems subject to monotonic loading. (74 references)

*K. Friedrich:* **Crazes and Shear Bands in Semi-Crystalline Thermoplastics.** The fifth chapter reviews the interaction between the microstructure of crystalline polymers and the craze or shear band formation at various temperature and loading conditions. (151 references)

*A. S. Argon, R. E. Cohen, O. Gebizlioglu, C. Schwier:* **Crazing in Block Copolymers and Blends.** The last chapter reports on the principles of polymer toughening by crazing in block copolymers and blends. (68 references)

Springer-Verlag
Berlin
Heidelberg
New York
Tokyo

W. Klöpffer

# Introduction to Polymer Spectroscopy

1984. 80 figures. XII, 190 pages
(Polymers. Properties and Applications, Volume 7)
ISBN 3-540-12850-6

**Contents:** General Introduction: Introduction. – Electronic Spectroscopy: ESCA in Polymer Spectroscopy. Absorption Spectroscopy in the Ultraviolet and Visible Regions. Fluorescence- and Phosphorescence Spectroscopy of Polymers. – Vibrational Spectroscopy: Vibrations Raman Spectroscopy. Infrared Spectroscopy of Polymers. – Spin-Resonance Spectroscopy: Principles of Spin-Resonance Spectroscopy. Electron-Spin-Resonance (ESR) Spectroscopy of Polymers. Nuclear Magnetic Resonance (NMR) Spectroscopy of Polymers. – Conclusion and Appendices. – Subject Index.

Polymer spectroscopy is the science of quantum resonance interactions of electromagnetic radiation with polymers. This monograph provides a survey on polymer spectroscopy which includes the most important resonant absorption and emission processes involving polymers and electromagnetic radiation. It is divided into three main parts covering electronic spectroscopy, vibrational spectroscopy, and spin-resonance spectroscopy. The methods treated include ESCA, UV and VIS absorption spectroscopy; fluorescence and phosphorescence spectroscopy; the Smekal-Raman effect and the spectroscopy derived from it; IR absorption including the far IR and FTIR; and ESR and NMR ($^1$H and $^{13}$C), including a short introduction to MAS. At the end of each chapter the strength of the method is discussed.

It is shown that only the interplay of several methods can give an adequate picture of a given polymer. Finally, the advantages and limits of spectroscopic methods are described and compared with the "classical" methods of polymer research. The subject is treated in an elementary fashion and much emphasis is given to the understanding of the basic processes. The main purpose of the book, which has taken shape from several series of lectures, is a didactic one. However, it will also be of use to experts who wish to learn something about the neighboring field of polymer spectroscopy.

The monograph is of interest to polymer chemists, polymer physicists, analytical chemists, plastics technologists, and spectroscopists in universities and industry.

Springer-Verlag
Berlin
Heidelberg
New York
Tokyo

**RETURN** **CHEMISTRY LIBRARY** 6263
**TO** ➡ 100 Hildebrand Hall  642-375